Handbook of Materials Handling

Compiled by the

TRANSPORTFORSKNINGSKOMMISSIONEN
(SWEDISH TRANSPORT RESEARCH COMMISSION)

Translator:
R. G. T. LINDKVIST, F.I.L., M.T.G.

Translation Editor:
RODERICK ROBINSON
Editor, *Materials Handling News*

Technical Editor:
GREGOR LUNDESJO
Rolatruc Limited

ELLIS HORWOOD LIMITED
Publishers · Chichester

Halsted Press: a division of
JOHN WILEY & SONS
New York · Chichester · Brisbane · Toronto

First published in 1985 by

ELLIS HORWOOD LIMITED
Market Cross House, Cooper Street, Chichester,
West Sussex PO19 1EB, England

Distributors:

Australia, New Zealand, South-east Asia:
Jacaranda-Wiley Ltd., Jacaranda Press
JOHN WILEY & SONS INC.
GPO Box 859, Brisbane, Queensland 4001, Australia

Canada:
JOHN WILEY & SONS CANADA LIMITED
22 Worcester Road, Rexdale, Ontario, Canada

Europe, Africa:
JOHN WILEY & SONS LIMITED
Baffins Lane, Chichester, West Sussex, England

North and South America and the rest of the world:
JOHN WILEY & SONS INC.
605 Third Avenue, New York, NY 10158, USA

© 1985 Ellis Horwood Limited

British Library Cataloguing in Publication Data
Handbook of materials handling. —
(Ellis Horwood series in industrial technology)
1. Materials handling — Equipment and supplies
I. Robinson, Roderick II. Lundesjo, Gregor
III. Transportforskningskommissionen
IV. Material hantering. *English*
621.8'6   TS180.3

ISBN 0-85312-574-0 (Ellis Horwood Limited)
ISBN 0-470-20098-7 (Halsted Press)
Printed in Great Britain by R.J. Acford, Chichester.

Foreword

Some years ago, the attention of TFK's materials handling committee was drawn to the need for a modern, comprehensive manual covering materials handling.

Prior to that, the most recent standard work — out of print for a long time — was published by AB Bygg- och Transportekonomi (BT) in 1967 and inevitably had become obsolete.

It was proposed, therefore, that TFK produce a new manual. The intention was to create a standard work which, as well as showing ways of choosing materials handling equipment, would give a broad survey of the various types available, their characteristics, applications and economics.

TFK contacted various suppliers to obtain backing for the project. The upshot was an agreement between TFK and BT to produce the manual jointly, assisted by additional finance from the Swedish Technical Development Board.

A management team was formed under the chairmanship of Colonel Erik Bodin, of the Swedish Defence Material Administration, to control the project, while an editorial committee headed by Mr. Björn Ljungström, Research Manager, TFK, as project manager was charged with writing and editing the contents.

The other members of the management team and/or of the editorial committee were:

Mr. Bertil Carlsson, Head of Section, Swedish State Railways
Mr. Tomas Bergling, Research Engineer, TFK
Mr. Anders Denell, Transport Specialist, Agro Shipping Ltd.
Mr. Bengt Gustavsson, Research Engineer, TFK
Mr. Kjell-Åke Johansson, Transport Specialist, Saab-Scania Ltd.
Mr. Birger Järnek, Transport Specialist, NK-Åhléns Ltd.
Mr. Gerhard Lundell, Sales Engineer, BT Ltd.
Dr. Kaj Ringsberg, President, ILAB Ltd.

To verify statements made in the text and to obtain first-hand illustrations, the editorial committee contacted a great number of equipment suppliers. The drawings are by Mr. Per Forsell and Mr. Hans Strand, and the layout by Mr. Lars Petterson, BT Ltd.

TFK wishes to express its gratitude to all those who, by their contribution of painstaking work, have made this manual possible.

Karl-Lennart Bång
TFK

Introduction

Materials handling and transport, i.e. the flow of materials in their totality, represent a large and growing part of the costs incurred by society and industry. Many judge this to be the area where the greatest opportunities for rationalisation are to be found.

This makes it more and more important for management to scrutinise and develop existing materials handling systems and to produce adequate data as a basis for new investment.

Nevertheless it is evident that acquisition of materials handling equipment frequently is not preceded by a systematic analysis of the requirements.

Industry's technical managers are primarily concerned with production and marketing. Materials handling matters are put aside, thus increasing the risk of investment mistakes.

A contributory factor may be that there is insufficient knowledge of the benefits offered by materials handling technology or that the knowledge is not submitted at the right decision-taking management level. This manual attempts to remedy this to some extent and to provide a comprehensive survey of the possibilities presented by materials handling technology. Its purpose is generally to heighten awareness of materials handling technology and to provide a sounder basis for decisions on modification or on new investment.

The manual addresses itself to a broad spectrum of manufacturing companies and covers a variety of mixed materials storage and handling techniques. It is also aimed at organisations operating terminal installations and warehouses such as lorry terminals, port installations, wholesale depots, retail stockrooms, distribution centres, etc. For these organisations, the manual may serve a number of purposes, first as a general guide and an aid to decision-making on modifying or replanning internal transport systems, and secondly as instruction material for company training schemes.

The manual is also suitable for instruction at most levels of public education. It presumes no detailed previous knowledge among readers.

There are three parts:

Part 1 advises and instructs on the choice of materials handling equipment. It includes sections on financial calculation, and a systematic method of selecting materials handling equipment is described.

Part 2 describes the majority of equipment available for internal transport and handling, primarily of general cargo and goods on pallets, and auxiliary handling equipment is also reviewed. The equipment's construction, characteristics and application are described. Most chapters also contain tables, diagrams or graphs specifying typical equipment. Part 2 is divided into five parts: unrestricted, area restricted, line restricted and position restricted materials handling equipment, and auxiliary equipment. Each part of the manual is distinctively colour-coded to help the reader.

Part 3 reports on how six Swedish companies solved their materials handling problems. The equipment chosen for different tasks within the context of a total system is described and reasons are given for the choices.

Contents

Preface to English translation

Anyone who is seriously interested in the practice of efficient handling and storage techniques has to acknowledge sooner or later the enormous contribution the Swedes have made to the field. In my case, the discoveries began first-hand with a visit to Sweden in 1972. Although I had been writing about materials handling for some years previously, it was a professional delight to see theories turned into practice, to inspect humane shop floors and to pass by hardware that appeared to glitter with industrial virtues. On all subsequent visits, there were equally important lessons to be learned and so I was pleased to accept the invitation to make my own small act of recognition by helping with the translation of this book.

As far as I can recall, no other book on materials handling has attempted quite the same brief as this one. The pictures dominate the text because it has to be recognised that there is still a need to sell materials handling to managers who are unaware of what can be achieved. Pictures do this more immediately than written argument, even though,

of course, the next stage in the process of persuasion consists of cold hard figures which quantify the savings in effort, time and specialised expenditure. *Materials Handling* provides the figures also.

Although this book has Swedish origins it is meant for international consumption. But the Swedes, being Swedes, have gone further than most other industrialised countries, with the result that they have already evolved technical jargon for a couple of items (and one very interesting concept) not in use elsewhere. Toby Lindkvist (who has shouldered by far the lion's share of the work) and I have done our best to integrate these details with the other more established material.

I certainly enjoyed having to re-scrutinise Scandinavian methods. I hope all English-speaking readers will share this enjoyment as well as the instruction the book contains.

Roderick Robinson
Editor, *Materials Handling News*

Methods

PART 1

Terms and abbreviations

Materials flow - equipment

Equipment selection routines

Calculation methods and forms of financing

British Standards

Terms and abbreviations

Rapid technical developments in the materials handling industry make compilation of an up-to-date authoritative terminology difficult, if not impossible, to achieve. This chapter presents some of the terms and abbreviations most commonly used in materials handling. They lay no claim to being definitive. A more comprehensive Swedish terminology is to be found in "Transportordlista", the transport vocabulary published by TFK in 1981.

Other authoritative terminologies are covered in a list of current technical publications and are referred to throughout this manual.

General terms

Batched goods Goods which can be handled in batches without a special load carrier.
Note: Examples of batched goods are wooden boards, steel rods, logs and wire coils.

Bulk material Material in an unpacked state.
Note: E.g., ORE, GRAIN, OIL, SLAG.

Charged pallet ratio The proportion of goods, in a given quantity, that have been loaded onto pallets.

Combined transport The use of several means of transport.
Note: Usually combined road and railway operations.

Distribution Measures connected with transferring merchandise from the producer to the consumer.
Note: In addition to transport, storage and handling "distribution" also includes marketing, order processing and payment.

External transport Transport between different companies, plants, etc. See also 'Internal transport".

General cargo A group of goods items which cannot be regarded as a batch.
Note: Individual haulage organizations generally define limits for what they regard as general cargo. Such limits could range from less than a full wagonload (rail haulage) to less than one ton (road haulage).

Goods and materials Objects considered in terms of transport, transfer or with reference to space requirements, etc.
Note: Includes all physical objects transported except people.

Handling Transfer of goods for the purpose of bringing them to the correct position for a subsequent operation.
Note: E.g., CLAMPING, LIFTING, TURNING, LOWERING, LOADING, UNLOADING.

Integrated transport Consecutive transport activities in which conveyors, load carriers, terminals and storage installations, etc., have been adapted in conformity with a uniform principle.
Note: Normally means CONTINUOUS HANDLING of the goods.

Internal transport Transport within a company, plant, etc.
Note: E.g., transport during the production process, between departments or between adjacent installations. Normally handled by the company itself.

Line restricted equipment Equipment guided along a predetermined transport route.
Note: AREA RESTRICTED and POSITION RESTRICTED (or FIXED) equipment may also be distinguished.

Load Mass (goods or people) constituting a load on a vehicle transport device, etc.

Load carrier A device to facilitate the carrying and securing of a load.
Note: Load carriers are normally standardized (e.g., pallets, containers, demountables, etc.).

Local transport Transport within a locality (municipality).
Note: Normally the locality (municipality) where the transport company is situated.

Logistics Integrated control and physical transfer of raw materials, and semi-finished and finished products from a supplier to consumers via raw materials store, processing and finished products store. Also called 'MATERIALS ADMINISTRATION' (MA).

Long-distance transport
Transport between regions.
Note 1: Long-distance transport may be carried out directly between supplier and receiver or via a terminal where INTERMEDIATE HANDLING of the goods is involved.
Note 2: It has been suggested that transport covering greater distances than 100 km be called 'long-distance transport'.
Synonym: INTER-REGIONAL TRANSPORT.

Materials
See 'Goods and materials'.

Materials administration
See 'Logistics' and 'Resources administration'.

Picking rate
A key figure indicating the rate at which a pallet load is completely picked. For example, a picking rate of 5 indicates that 5 visits by the order picker accounts for an entire pallet load. Used to calculate the turnover of empty pallets in warehouses.

Multiple transport
Involves several goods movements, with intermediate goods handling operations.

Order
A customer's instruction for something to be supplied. An order may include a number of different order lines (product numbers) each one of which in its turn may consist of one or several goods items.

Pallet
A load carrier, normally having a rectangular, standardized load area. Consists of one face with supports, or two faces with spacer blocks, designed to admit the tines of a forked handling device.

Pallet charging
Positioning packages, cartons, sacks, barrels, etc., i.e., goods items, on a pallet.
Note: Not to be confused with "palletising", a term which should ideally be reserved for the introduction of a pallet system. Removing goods items from a pallet is termed 'PICKING'.

Pallet charging pattern
A volumetric pattern for building up smaller items into a pallet load.

Pallet loading
Loading of pallets on a transport device, etc.
Note: Hence, also, 'PALLET UN-LOADING'.

Parcel
A self-supporting goods unit.
Note: It may be, for instance, a BUNDLE, PACKAGE, BOX, CAN, BARREL, PALLET, BOTTLE, or other units, the dimensional limits of which may be difficult to specify.

Regional transport
Transport within a region (county). (Cf. 'Local transport'.)

Resources administration
A collective term covering capital and materials administration.

Sorting
A transport terminal activity by which goods are divided into groups. Goods in each group should then have the same intermediate or permanent destination, mode of transport and/or delivery route.

Store
An installation for storing materials, plant components or equipment (cf. 'Warehouse').
Synonym: Store-room.

Stowage
Handling for the purpose of positioning and, where required, securing goods in the space intended.

System tranport
Integrated transport operated according to predetermined routines.
Note 1: E.g., according to a time-table.
Note 2: 'RETURN TRANSPORT' is a term often employed in connection with goods being moved in the opposite direction to that primarily employed in system transport.

Terminal
An installation where goods are loaded/unloaded and passengers embarked/disembarked.
Note: E.g., GOODS TERMINAL, BUS TERMINAL, AIR TERMINAL.

Transfer
Varying the position of objects.
Note: A transfer may be manual (e.g., CARRYING, ROLLING), or may be effected by means of aids (e.g., HOISTING, CARTING).

Transhipment
An overall term covering loading, unloading and marshalling.

Transport
Moving goods (or people); also associated handling and storage operations.

Transport system
Physically or administratively co-ordinated means of transport with associated installations.

Unit load
A material in a packed state. Frequently, a standardized size transport unit.
Note: Normally a large goods unit comprising several sub-units positioned on or in a standardized load carrier such as a pallet, container or platform for the purpose of being handled as a unit during a complete transport process. However, some unit loads do not require a load carrier, e.g., TIMBER BUNDLES, PULP BALES.

Warehouse
An installation for storing products during long gaps between production stages or for storing finished products. (Cf. 'Store').
Synonym: DEPOT.

Abbreviations

ADR Accord Européen pour le Transport de Merchandises Dangereuses par Route (the European Agreement on Road Transport of Dangerous Goods).

ASF Arbetarskyddsfonden (the Swedish Industrial Safety Fund).

ASS Arbetarskyddsstyrelsen (the Swedish Industrial Safety Board).

AV Arbetarskyddsverket (the Swedish Industrial Safety Agency).

FEM Fédération Européene de la Manutention. A European authorities for transport investigation and research. materials handling equipment.

ICHCA International Cargo Handling Co-ordination Association.

IKH IVA's Crane and Lift Commission.

ISO International Standards Organization.

IVA Ingenjörsvetenskapsakademien (the Swedish Academy of Engineering Science).

NKTF Nordiska Kommittén för Transportekonomisk Forskning (the Nordic Committee for Research on Transport Economy).

SIS Standardiseringskommissionen i Sverige (the Standards Commission of Sweden).

SMS Sveriges Mekanstandardisering (the Swedish Medical Engineering Standards Organization).

STU Styrelsen för Teknisk Utveckling (the Swedish Technical Development Board).

TFD Transportforskningsdelegationen (the Swedish Transport Research Delegation). A government research council (under the Ministry of Transport).

TFK Transportforskningskommissionen (the Swedish Transport Research Commision). An association of business firms and authorities for transport investigation and research.

TTF Transporttekniska Föreningen (the Swedish Transport Engineering Association).

TYA Transportfackens Yrkes- och Arbetsmiljönämnd (the Swedish Transport Industry's Occupational and Working Environment Committee).

VTI Statens Väg- och Trafikinstitut (the Swedish Road and Traffic Institute).

List of glossaries

In addition to the transport glossary, 'Transportordlista' published by TFK in 1981, current Swedish transport terms may be found in the publications listed below.

Chapter	Publication
Trucks, carts and trolleys	Truck and trolley terminology. SMS 27 80 (68 terms).
Overhead travelling cranes Transporters	Cranes, terms. IKH 1.30.01. Loading devices, terms. IKH 1.50.02. Illustrated terminology of heavy lifting tackle and cranes. FEM, Mekanförbundet 61 103.
Stacker cranes	Stacker cranes. ASS Directions No. 134 (1978).
Roller conveyors Wheel conveyors Belt conveyors Slat conveyors	Conveyor terminology and symbols. SMS report 987. FEM illustrated terminology of continuous conveyors. Mekanförbundet 78 110.
Vertical conveyors	Lifts, terminology. SIS 76 35 00. Illustrated terminology of lifts and escalators. FEM, Mekanförbundet. Order No. 61 106.
Load carriers	Pallets, terminology. SIS 84 10 00. Demountable bodies. Swedish standard specification. SMS 3038 Containers. SIS 842101. Container marking codes. SIS 84 21 07. Glossary, terms and definitions used in container service. (Four languages.) SJ. Tanks for liquids and gases (Swedish–English). SIS 84 21 08.
Warehousing lay-outs	Pallet racks. SMS 2240. Shelf racks. SMS 2241.

Materials flow - equipment

Equipment alternatives for the internal materials flow functions

The materials handling concept in its widest meaning covers all physical handling and transport of 'materials' in the entire distribution chain, from raw materials production to finished product consumption.

This manual confines itself to a smaller area and deals principally with internal handling. It therefore describes equipment needed for internal materials flow, i.e., the flow which begins and ends at the points where it connects with external transport.

Internal materials flow provides a company with the greatest scope for choice of equipment, even to the extent of loading and unloading external means of transport. Beyond this, i.e., external transport handling and vehicle design, it is more difficult to exert an influence.

This chapter examines the choice of materials handling equipment relating to the various main functions of internal materials flow. The different solutions for each main function are discussed together with possible equipment alternatives.

Internal materials flow in a manufacturing industry may be divided roughly into five main functions:

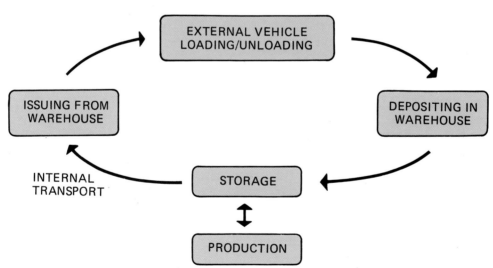

Main functions of internal materials flow.

A sixth function – internal transport – links the handling or processing functions. If the production function is excluded, the diagram may also be applied to a trading or transport company with terminal or warehousing operations.

Loading and unloading external transport vehicles

Choice is governed by the following four factors:

- The nature of the goods (load unit characteristics, packaging, fragility, etc.).
- The materials flow — its scope and its timing.
- The need to accommodate other internal flow systems.
- The need to accommodate other external flow systems.

Where optimization of the total materials flow (external and internal) is sought, it is obvious that the choice of packaging method, load unit characteristics and loading/unloading methods are greatly influenced by the need to accommodate other aspects of the materials flow.

In many cases, however, packaging, load and materials flow timing may be predetermined.

Other factors potentially important to loading/unloading methods are:

- Type and size of external transport vehicle together with its access facilities and associated handling equipment.
- Building design, the existence of loading bays, etc.

Suitable equipment for loading and unloading is listed below on the basis of material handled and loading/unloading method.

Loading and unloading equipment		
Goods type	Loading bay handling (from the side: lorries and trains; from the back: lorries)	Ground level handling (from the side or the rear)
Mixed general cargo (non-pallet loads, i.e., cardboard boxes, tins, cases, carpet rolls, wheelbarrows, bicycles, etc.).	Manual handling. Sack truck. Manual handling supplemented by extendable belt conveyor reaching into the vehicle (suitable for cardboard boxes, sacks and cases). Lorry-mounted steel belt conveyor or slat conveyor, Manual handling + pallet + hand or powered pallet truck.	Less usual for mixed general cargo except as part of distribution chain. In the latter case, the sack truck is most commonly used. There are also special wheeled pallet cages.
Roll pallets.	Manual handling. Powered stacker or counterbalanced truck.	Manual handling combined with lorry tail-lift.
Pallet loads.	Hand pallet truck (normally cannot handle large slatted wooden crates). Counterbalanced forklift + hand pallet truck. Low headroom counterbalanced truck. Chain or roller conveyor on loading bay and vehicle (for highly mechanized operations).	Counterbalanced truck. Reach truck (requires fairly even surface). Vehicle mounted crane with fork attachment. Hand pallet truck + tail-lift. Special crane with fork attachment for simultaneous unloading of 10—20 pallets.
Lengthy loads. Heavy loads.	Less usual.	Counterbalanced truck with sufficient fork width to accept lengthy loads. Sideloader (for lengthy loads). Gantry crane or overhead travelling crane (pre-supposes possibility of overhead access to tarpaulin-covered lorries or containers, or railway wagons). Vehicle mounted crane.
Container.	Less usual. It is possible to fit the container with wheels at each corner and roll it on to the bay (although this pre-supposes lifting facilities).	Forklift with spreader attachment. Sideloader. Straddle carrier. Mobile lifting frame. Gantry crane. Vehicle mounted sideloader with extendable outriggers. Separate hydraulic legs.
Demountable body.	As for container above.	Vehicle mounted hydraulic/mechanical or pneumatic lifting equipment.

15

Mixed general cargo may comprise anything from cardboard boxes and cases to carpet rolls and unpacked goods such as wheelbarrows and bicycles, etc. It is handled mainly through the distribution services run by freight forwarding companies and the railways. Loading bays are desirable to enable personnel to board the lorry or rail wagon to stow or unload the goods.

Sack trucks (see 'Carts and trolleys') are commonly used to handle goods packed in cardboard boxes

Mixed general cargo is most commonly loaded/unloaded from a bay. This involves considerable manual labour in combination with the old stalwart, the sack truck.

or cases. The goods may also be placed on a pallet and moved on board by hand pallet truck, subsequently to be stowed by hand. Sometimes, extendable belt conveyors, or wheel or roller conveyors are used to move goods into the vehicle where they are often stowed manually.

Some lorries are equipped with steel belt or roller conveyors. This arrangement, however, requires similar equipment at the loading bay.

Mixed cargo sometimes has to be unloaded where there is no loading bay. The most common solution is a tail-lift in combination with a hand pallet truck or sack truck. Alternatively, a special roll pallet may be used. Vehicles which offer access from the side demand a much more flexible unloading system, especially if goods in cardboard boxes are involved.

Pallet loads can be handled from ground level or loading bay.

Unloading from ground level is simplest and easiest where access to the vehicle is from the side. Curtainsided lorries provide this facility; a less common alternative is where the whole side of the lorry opens like a door.

Rail wagons with fully opening sides are used mainly for moving materials from loading bays.

Unloading from ground level by counterbalanced truck is a quick and flexible method where cargo is accessible from the sides.

Counterbalanced trucks are commonly used to unload from ground level. Reach trucks are also used in some cases, e.g., in indoor unloading operations which include placing the materials in storage.

The choice of equipment is often determined by the other uses to which it may be put as, in most firms, loading and unloading are not a continuous daylong process.

This means that loaders, etc., fitted with fork sets can also be used.

To deliver pallet loads where the customer has no suitable handling equipment, a vehicle mounted crane with pallet forks may be used. A combination of tail-lift and hand pallet truck may be suitable with more lightweight goods.

Powered pallet truck moving pallet loads onto loading bay.

driving onto the vehicle.

Low-headroom counterbalanced trucks can be driven into lorries and containers.

Where high levels of pallet unloading are required, powered roller conveyors and chain conveyors may be used. The best throughputs are achieved where the bay is also equipped with a conveyor matching that mounted on the vehicle. Alternatively, the vehicle-mounted conveyor may be used to advance the pallets, making them accessible at the rear to a counterbalanced truck or reach truck (e.g. where the vehicle has an insulated body).

For very high throughputs, special cranes with fork attachments can handle large numbers of pallet loads simultaneously. The crane shown can load an entire vehicle in two cycles.

Discharging pallet loads at a loading bay is normally not as easy and flexible as working from ground level. It can be carried out from the rear, from the side or from both directions at the same time. The lorry or rail wagon inevitably occupies a proportion of the bay, so that maximum handling capacity is limited by available bay length. When unloading from the rear, the handling equipment should, for maximum efficiency, be capable of being driven onto the lorry deck or into the rail wagon. Otherwise, one type of handling equipment will be necessary to move the goods to the edge of the bay where a different type will take over. This combination may be a hand pallet truck on the vehicle and a counterbalanced truck on the loading bay.

Typical aids to removing pallet loads from a bay are:

Hand pallet trucks, most commonly used at general cargo distribution centres. Note: inclines and sills must be kept to a minimum. A powered pallet truck may give slightly better results.

Powered stackers, either ride-on or pedestrian. These are lightweight and compact, suitable for

Pallet loads of frozen products are moved into and out of this semi-trailer by built-in chain conveyors which mate with a short length of chain conveyor on the loading bay. The goods are then collected by reach trucks used for picking and depositing stock in the adjoining cold-store.

17

When loading and unloading lengthy or heavy loads a gantry crane (overhead travelling crane) is commonly used and is standard equipment in steel stockholding. Where an overhead travelling crane is not available, a heavy vehicle mounted crane provides an alternative. Unloading by truck from the side is also possible. To ensure safe handling, however, the fork spread must be wide enough. Sideloaders are, of course, specially designed to handle lengthy loads.

At this steel stockholder, lorries are loaded using overhead travelling cranes. (Gantry cranes are the more normal outdoor alternative).

For unloading complete containers or demountable bodies a number of techniques are employed. (For demountables, see 'Load carriers' where various unloading methods are described):

A vehicle mounted sideloader with outriggers. The container is unloaded sideways. The equipment is comparatively heavy, reduces the payload and is fairly slow, so it is suitable primarily for infrequent handling activities, e.g., in a distribution operation.

A counterbalanced truck with a spreader attachment. Full containers must be lifted using the twistlocks. Full size containers should be handled on the forks only when empty although smaller containers and demountable bodies may be handled this way.

Separate hydraulic legs. These are secured to the corner fittings on the container which can then be lifted and lowered. Suitable for infrequent handling.

The gantry crane is the most common lifting aid at terminals. It can straddle a number of railway tracks and lorry lanes as well as special storage areas, serving both railway wagons and lorries.

Large counterbalanced truck equipped with a spreader attachment, handling a 20 ft ISO container.

Warehouse input procedures

The warehouse is a very cost intensive part of a transport system because of the low rate of throughput and the resultant tying-up of large amounts of capital. An appropriate internal layout and suitable equipment are essential for optimum handling results. The input side of the warehouse may incorporate the following operations:

Goods input ▷

Quantity control ◁

Quality control ◁

Repacking ☐

Labelling ◁

Accumulation in buffer stock ▷

Goods collection ☐

Removal to warehouse ⬇

Examples of warehouse input procedures.

Examples of internal transport equipment used to supply the warehouse		
	Removals to warehouse	
Internal transport equipment	**Pallet loads**	**Smaller storage units**
Automatically guided vehicles	●	
Roller conveyors	●	
Order pickers		●
Stackers	●	
Counterbalanced trucks	●	
Pedestrian and ride-on powered pallet trucks	●	●
Ride-on stackers	●	
Reach trucks	●	

The above operations are also known as 'goods reception'.

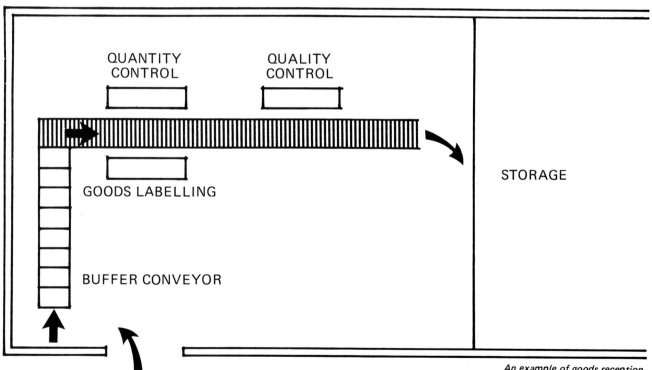

An example of goods reception.

The labour required for stock input depends on whether incoming goods are on pallets and may be moved directly to the warehouse or whether they consist of smaller units which need to be built up into larger packs. Where pallet loads are moved intact from an external vehicle to the warehouse, labelling is often done manually as the pallets pass a convenient point.

Quality and quantity checks are often based on random sampling.

A powered or unpowered roller conveyor is often used to accumulate goods on either side of the receiving station.

Warehousing

(See also 'Warehouse storage systems'.)

When designing a warehouse the building costs should be compared against the costs of the storage system. When comparing two alternative building designs a more expensive transport/handling layout may permit the reduction of building costs.

Because good space utilization is essential in warehouses, this makes demands on:

- type of storage and racking height.
- lifting height and width of handling equipment.
- layout.

When using existing premises for a warehouse, it is important to consider the floor utilization (as it relates to different types of storage systems) as well as space utilization. Thus, less floor area is required when pallets are stored with their short sides facing the aisle, but accessibility is impeded, e.g., when picking from pallets.

Another essential factor is the choice of **random** or **fixed locations** for the warehouse storage units. 'Fixed locations' means that each unit is given a specific location in the warehouse, with buffer stations close by. 'Random locations' means that the unit may be placed anywhere in the warehouse where there is an empty space; as a result buffer stations may be remote.

While fixed locations result in a comparatively low level of space utilization, they do provide spare capacity for random location storage. Random locations make greater demands on stock accounting and control systems.

The term **'small goods'** refers to units smaller than a pallet load. Such goods may be stored on shelves or in binning. Larger units such as cases and cardboard boxes are stored on shelves. Small parts such as screws are stored in bins.

Handling equipment for small goods is dealt with in the next section, 'Warehouse output procedures'.

A **pallet warehouse** may be based on one of three storage principles: pallet racking, live storage and block stacking (see also 'Warehouse storage systems'). Pallet racking and block stacking may be designed for long or short side handling. In live storage the pallet's orientation will provide the basis for its design, i.e., the direction in which the pallet moves along the conveyor.

With pallet racking any pallet may be retrieved. Live storage is based on the 'first in—first out' principle. Block stacking is suitable principally for large volume applications involving only a few product lines which, themselves, have a long shelf-life; this is because the pallet stacked first is the last one retrieved. Where pallet loads cannot be stacked directly on top of one another in block stacking, drive-in racks with supporting rails for the pallet runners are used.

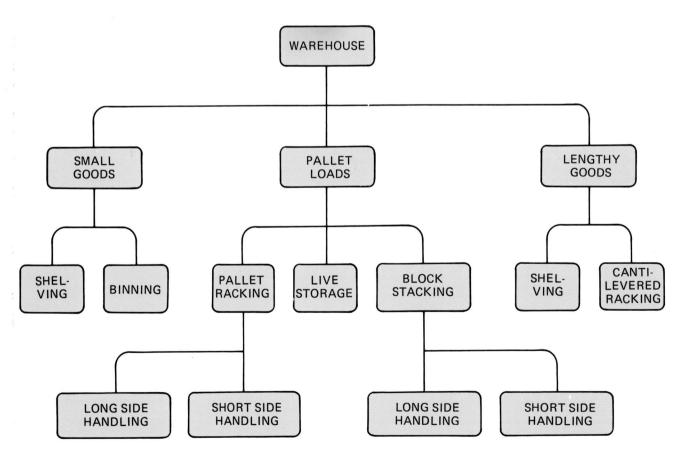

An example of storage systems in a multi-type goods warehouse.

Examples of internal transport equipment for warehouse handling of pallet loads					
Internal transport equipment	**Pallet racking**		**Live storage**	**Block stacking**	
	Long side handling	**Short side handling**		**Long side handling**	**Short side handling**
Hand stacker		●			
Pedestrian or ride-on powered pallet truck		●			
Pedestrian stacker		●	●		●
Ride-on stacker		●	●		●
Counterbalanced truck	●	●	●	●	●
Stacking with powered lift		●	●		●
Reach truck	●	●	●	●	●
Narrow-aisle truck	●	●			
Order picking truck	●	●	●		
Sideloader	●	●			
Stacker crane	●	●			
Stacking overhead travelling crane	●	●			

With lengthy goods special storage arrangements and handling equipment are necessary. Such goods include steel bars, sawn timber and plastic piping and may be stored on cantilevered racking or shelves.

Four-way trucks, sideloaders, stacker cranes and overhead travelling cranes may be suitable as internal transport equipment.

Warehouse output procedures

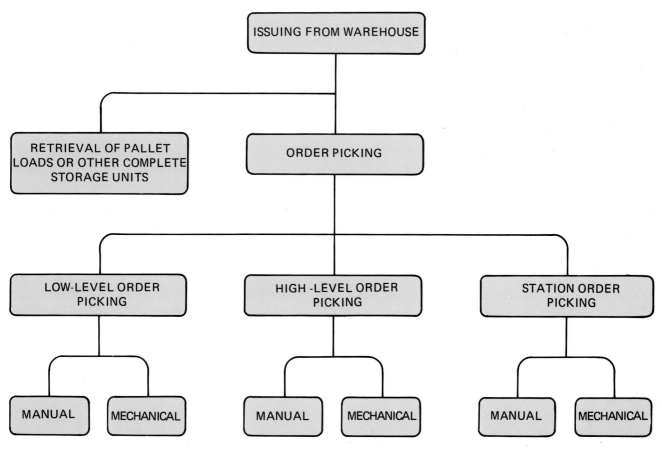

Warehouse retrieval principles.

The retrieval of a pallet load or other unit for direct delivery may involve the following operations:

Collecting the goods from the warehouse ☐

Moving the goods to the dispatch station ⬇

Accumulation ▷

Delivery check ◁

Labelling ◁

Accumulation ▷

An example of the operational stages when retrieving a pallet load or other unit from a warehouse for direct delivery.

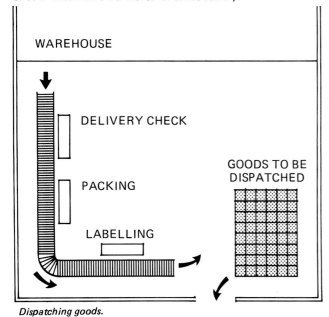

WAREHOUSE

DELIVERY CHECK

PACKING

GOODS TO BE DISPATCHED

LABELLING

Dispatching goods.

A warehouse arranged for **order picking** may be defined as an area where materials are stored and retrieved mainly in units smaller than those used for basic storage. Order picking warehouses are classified according to the following factors:

- Picking method
- Buffer zone
- Retrieval principle
- Pre-delivery operations

Three order picking methods are possible: low-level, high-level and station picking. Low-level and high-level picking are based on the principle of 'man to goods'; station picking represents 'goods to man'.

The position of the **buffer zone** in a warehouse may vary. It could be:

- A remote buffer not directly connected with the picking zone.
- A buffer adjacent to the picking zone but not directly accessible to the picker.
- A buffer located in the picking zone itself.

The **retrieval procedure** in an order picking warehouse depends on whether picking involves the entire goods range or merely part of it, also whether one or several orders are picked simultaneously.

Pre-delivery operations consist of sorting and packing which may be done in direct conjunction with picking (the pick/pack principle) or in separate zones subsequent to picking. Pick/pack is normally employed for picking one order per visit and/or several order lines per order.

The goods dispatch and reception stations are frequently adjacent in order that the same handling equipment employed for moving pallet loads can be utilized.

To sum up, manual order picking may be performed according to the table below.

Classification of manual order picking methods												
Picking method	Retrieval principle				Buffer zone			Pre-delivery operations				
	Product range		Number of orders					Sorting		Packing		
	Full	Partial	One	Several	Remote	Adjacent	In the Picking zone	Direct when picking	In a separate zone	Direct when picking	In a separate zone
High/low level method	●	●	●	●	●	●	●	●	●	●	●
Station picking	●			●	●			●			●

An example of classification of manual picking methods.

22

Mechanical order picking is used less extensively but may become increasingly popular as labour costs rise.

To conclude, the various order picking methods have the following characteristics:

- Low-level order picking, where the operator works from floor-level, is used principally where there is a high turnover of goods and multi-product orders.
- High-level order picking takes place at elevated storage levels and where the range of stored goods is large but the buffer requirements are small.

- In station picking the goods are transported to the picking station as storage units. Station picking is employed mainly where the range of goods is small and several orders are picked at the same time.
- For low or high-level picking from pallets, the long side handling storage technique offers best access. However, it will sometimes result in less satisfactory warehouse utilization.

Live storage can be used in conjunction with order picking.

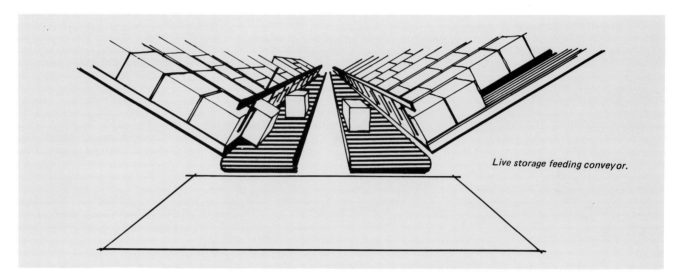

Live storage feeding conveyor.

The goods are live stored in cardboard boxes at right angles to a conveyor. This arrangement permits fully mechanized picking. During retrieval, the restraint on the lowest box is released, and the box slides on to the conveyor belt and passes to the dispatch station. This system may also be used with advantage in manual order picking.

A pick/pack machine with grippers for bottle crates.

Production

Engineering production can be divided into manufacture and assembly. In manufacture, the handling equipment is subordinate to the other factors (materials, labour, machinery) controlling the design of the production line. In assembly operations, mechanical handling is the dominant factor.

Pure **line assembly**, with closely controlled production and linear flow, requires equipment of great reliability and high rate of handling in those cases where station times are short.

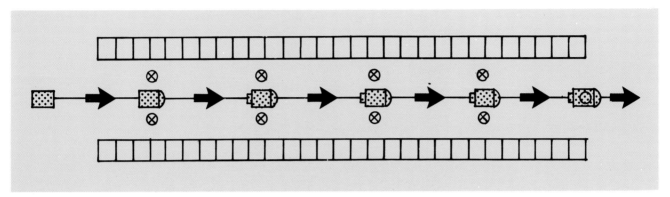

Line assembly

In **group assembly,** stations are fewer and station times are longer. Stations are often positioned to one side of the main line of the assembly flow. Mechanical handling equipment must thus be capable of moving in any direction.

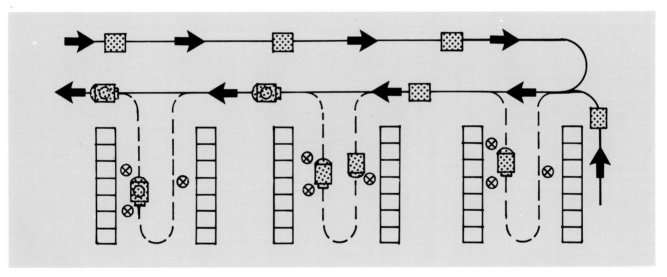

Group assembly

In assembly operations an object is often moved on mobile work platforms which act as jigs or fixtures.

Examples of internal handling equipment for assembly operations		
Internal handling equipment	Line assembly	Group assembly
Roller conveyors	●	
Belt conveyors	●	
Chain conveyors	●	
Slat conveyors	●	
Overhead travelling cranes	●	●
Telpher system	●	●
Automatically guided vehicles	●	●
Air-cushion systems		●

Equipment selection routines

When choosing an internal mechanical handling system or acquiring new handling equipment the general methodology below may be used.

The diagram shows the steps involved, the guidance provided by this manual and at what stages of the evaluation contact should be made, for instance, with a supplier.

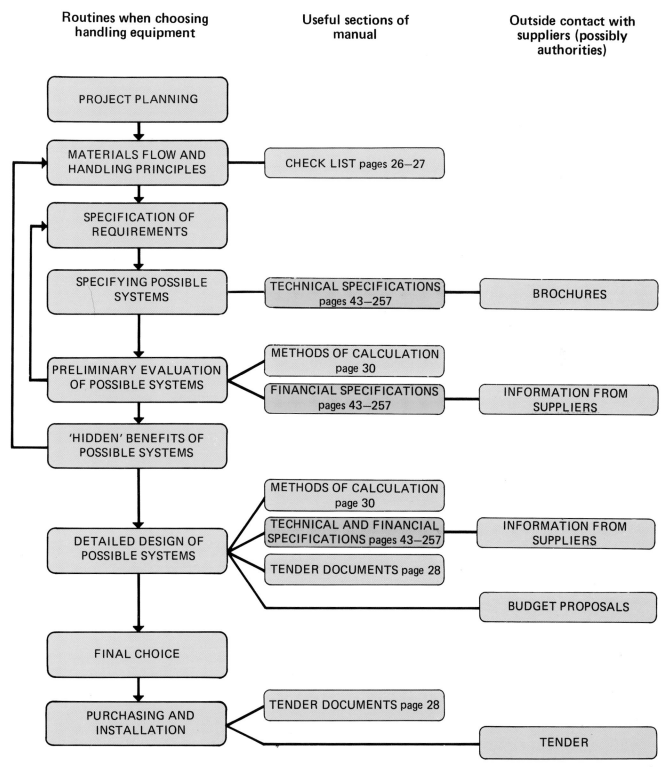

| Routines when choosing handling equipment | Useful sections of manual | Outside contact with suppliers (possibly authorities) |

PROJECT PLANNING

MATERIALS FLOW AND HANDLING PRINCIPLES — CHECK LIST pages 26—27

SPECIFICATION OF REQUIREMENTS

SPECIFYING POSSIBLE SYSTEMS — TECHNICAL SPECIFICATIONS pages 43—257 — BROCHURES

PRELIMINARY EVALUATION OF POSSIBLE SYSTEMS — METHODS OF CALCULATION page 30 / FINANCIAL SPECIFICATIONS pages 43—257 — INFORMATION FROM SUPPLIERS

'HIDDEN' BENEFITS OF POSSIBLE SYSTEMS

DETAILED DESIGN OF POSSIBLE SYSTEMS — METHODS OF CALCULATION page 30 / TECHNICAL AND FINANCIAL SPECIFICATIONS pages 43—257 — INFORMATION FROM SUPPLIERS / TENDER DOCUMENTS page 28 — BUDGET PROPOSALS

FINAL CHOICE

PURCHASING AND INSTALLATION — TENDER DOCUMENTS page 28 — TENDER

Project planning

When re-designing or modifying a transport system, project planning is necessary and should include the following tasks:

- Defining and delineating problems in respect of multiple transport and objects handled.
- Setting project objectives to clarify the expected results of the re-design/modification.
- Drawing up timetables and resources schedules to establish:
  - Project managers and contacts.
  - Investment by sub-operation.
  - Time required with respect to the personnel concerned; also time allocations for decision making by management, authorities, etc.
  - Project costs of company and outside personnel, distributed by sub-operation and covering the total project.

Goods flow and transport principles

Choosing transport equipment requires an analysis of the materials flow and transport principles. The extent of this analysis depends on:

- Cost level of internal transport system.
- Investments already made.

The analysis covers:

- Materials to be transported (size, weight, etc.).
- Materials flow (items per time unit).
- Work and buffer stations.
- Transport routes.
- Company based cost data.

Flow analysis and transport principles are arrived at by viewing the movement and treatment of the materials flowing through the system. In a warehouse these aspects may be studied by investigating the major warehouse operations such as reception, unpacking, storage, picking, packing and dispatch.

Suitable choice of equipment and improved administration and control may permit a reduction in the number of operations, for instance, in conformity with the diagram below.

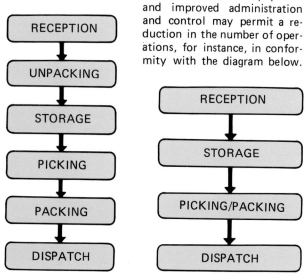

An example of redistribution operations in a warehouse.

To determine whether or not a redistribution is feasible the various activities are listed in a process analysis chart such as the one below:

Activity No.	Description of activities	Processing	Handling	Transfer	Storage	Inspection/hold
		○	□	⇨	△	▽
	Unloading from delivery vehicle	○	□	⇨	△	▽
1	Collecting goods	○	■	⇨	△	▽
2	Transfer to goods reception	○	□	➡	△	▽
3	Positioning goods	○	■	⇨	△	▽
	Goods reception	○	□	⇨	△	▽
4	Accumulating incoming goods	○	□	⇨	▲	▽
5	Quantity control	○	□	⇨	△	▼
6	Quality control	○	□	⇨	△	▼
7	Repacking	○	■	⇨	△	▽
8	Labelling	○	□	⇨	△	▼
9	Accumulation	○	□	⇨	▲	▽
	Storage input	○	□	⇨	△	▽
10	Collecting goods	○	■	⇨	△	▽
11	Transfer to warehouse	○	□	➡	△	▽
12	Depositing stock	○	■	⇨	▲	▽
13	Storage	○	□	⇨	△	▽
	Storage output	○	□	⇨	△	▽
14	Collecting goods	○	■	⇨	△	▽
15	Transfer to goods assembly station	○	□	➡	△	▽
16	Positioning goods	○	■	⇨	△	▽
17	Accumulation	○	□	⇨	▲	▽

	Goods assembly	○	□	⇨	△	▽
18	Picking	○	■	⇨	△	▽
19	Delivery checking	○	□	⇨	△	▼
20	Packing	○	■	⇨	△	▽
21	Labelling	○	□	⇨	△	▼
22	Accumulation	○	□	⇨	▲	▽

	Dispatch	○	□	⇨	△	▽
23	Collecting goods	○	■	⇨	△	▽
24	Transfer to delivery vehicle	○	□	⇨	△	▽
25	Positioning on delivery vehicle	○	■	⇨	△	▽

Process analysis chart describing a transport system.

Goods and production data may be summarized on a form intended to list factors affecting the design of the transport system (see arrangement below).

Unit type:	**Risk of damage to units:**
Unit weight: kg	
Size: Length: mm Width: mm Height: mm	**Load carrier requirements**
Position of the centre of gravity relative to the centre of the unit Length: % Width: % Height: %	

Unit flow average: maximum: units/day	
Transport system throughput time: minutes/unit	
Method requirements: ☐ Lifting ☐ Turning ☐ Rotating ☐ Stacking ☐ Picking	**Environmental requirements:** ☐ Noise ☐ Exhaust gases ☐ Vibration

Work station type:	**Buffer station type:**
Operation time: average: maximum: min/unit	**Accumulation time** average: maximum min/unit
Work area: m²	**Buffer area:** m²

Transport route: horizontal: m vertical: m incline: m (%) width: m	**Floor type:**
	Compression strength: N/m²
	Space limitations:

Personnel costs: SEK/month year	**Floor space costs:** SEK/m², year
Calculated interest: %	**Electricity costs:** SEK 0.01/kWh

A form for summarizing goods and production data, and operational and environmental requirements.

Specification of requirements

Each materials handling activity should be analyzed with respect to:

- Current requirements.
- Origins of these requirements.
- Degree of urgency involved.

Examples of requirements are horizontal and vertical transferability, load-carrying capacity, required space, and technical and financial life.

The materials handling system is subject to overriding requirements such as rate of throughput, buffer capacity, required personnel, self-monitoring and self-adjusting abilities, energy and service required, investment required, and capital and operating costs.

The same requirement may originate from **several sources**. These sources may be personnel, management, authorities or the community.

Specifying possible handling systems

Where **load carriers** are required, the analysis should examine alternative load carriers. Load carriers are required where the characteristics of the materials transported need improving with regard to ease of handling, transportability, accumulation, load security, and ease of accessibility for mechanical assembly.

When considering alternative unit loads, it is important to check whether the load carriers are available within the company and/or from commercial suppliers.

Transport equipment is examined in Part 2 of this manual. Load carriers may make special demands on the transport equipment, and this may limit the number of possible systems. Other load carriers may, therefore, be necessary.

When considering possible **transport systems**, the required number of transport units and load carriers should be calculated. Studies should also include transport capacity, buffer capability, route distances, vehicle speeds, station times and other characteristics.

Possible systems should be examined with regard to:

- Capacity.
- Flexibility.
- Reliability.
- Safety.
- Energy, service and labour requirements.
- Controllability.
- Feasibility of integration.
- Noise, exhaust fumes and other environmental factors.

Preliminary evaluation of possible systems

The range of possible transport systems now obtained from the preceding stage should then be evaluated in terms of finance. Calculations should cover capital requirements for investment, capital and operating costs, personnel costs, cost of floor space, and administration and management costs. Calculation methods are described in the chapter 'Calculation methods and forms of financing'.

Possible transport systems should be checked against both average and maximum materials flow. This will result in:

- Systems meeting the requirements completely.
- Systems not meeting all requirements but capable of being modified to the standard required.
- Systems which should be excluded because they do not meet the requirements.

The possible transport systems can then be ranked according to desirability.

'Hidden' benefits of possible systems

Such a ranking will only apply to static conditions. The ranking may be changed, however, if some systems provide increased flexibility or opportunities for new production arrangements and consequent variations in production layout and materials flow.

Before the possible systems are worked out in detail, therefore, the opportunities for improved production should be investigated. This may result in new systems and/or in changes in the original ranking.

Detailed formulation of possible systems

The best of the possible systems should be planned in detail before a final choice is made. To this end more detailed technical and financial data are necessary, e.g., costings from the manufacturers. The costs should be based on a specification which lists the equipment requirements as defined by the analyst.

The **tender documents** should be based on the equipment alternatives obtained from the preliminary evaluation. They should include:

- A brief description of the equipment's working environment.

- A description of the materials and the load carriers to be handled.
- The technical and environmental requirements of the equipment.
- The number of items of equipment required.
- An expansion of the costing to include price per item, delivery commitment, date of delivery, warranty period and extent of assembly work.
- A contact name in the customer's company.

Final choice

When detailed costings have been received, a final technical and financial analysis of the various possibilities should be carried out against the background of stated requirements. This assessment should take into account such matters as spare parts, time of delivery, personnel training, etc., offered by different suppliers.

The 'best' handling systems may then be selected. Firm quotation enquiries should be prepared for the purchase department.

Purchasing and installation

Tender documents should be prepared for the equipment which corresponds to the needs of the selected handling system.

In addition to meeting the requirements listed earlier, tenders should also include the suppliers' undertakings with respect to installation and commissioning.

Calculation methods and forms of financing

Cost concepts

Costs may be subdivided in different ways for different purposes. Their subdivision into **overheads** and **traceable costs** takes into consideration the causes of their origin. A traceable cost is linked entirely to an individual vehicle or equipment while an overhead is not dependent on any individual object but is shared among other objects.

Direct and indirect costs take account of whether or not a cost incurred can be directly charged to a product or some other cost-bearing unit.

Fixed and variable costs are separated on the basis of their dependence on the volume of production. In the short term, a fixed cost is independent of production volume, whilst a variable cost changes with it. Transport and materials handling practise is different and it is customary instead to differentiate between **fixed** costs and **operating** costs. A fixed cost is independent of equipment usage while an operating cost depends on transport distance and/or the equipment utilization period.

Calculating transport costs

The following facilities are normally required to provide transport services:

- Transport equipment and load carriers.
- Drivers or operators.
- Workshop.
- Storage and office premises.
- Administration and management.

Together, these facilities represent the transport costs and are divided into:

Fixed costs (SEK per year)

1. Capital costs (interest charges and depreciation).
2. Tax and insurance (equipment, operators, goods).
3. Wages including social security contributions (operators).
4. Sundry fixed costs (environmental costs, storage premises, etc.).

5. Administration (salaries, rent, materials, etc.).

Operating costs (SEK per hour or km)

6. Fuel and power (energy costs).
7. Maintenance (service, lubrication, vehicle cleaning, etc.).
8. Repairs (cost of unplanned repairs and, possibly, downtime).
9. Tyres, batteries, etc.

Capital costs are divided into depreciation and interest on capital tied up in equipment. Interest may be regarded as the 'price' paid for using the amount of money needed for an acquisition. Interest level is normally determined centrally in the company. 'Calculated depreciation' means that the acquisition expenditure is spread over the economic life of the investment. Thus, each year of usage is charged with a cost corresponding to the utilized share of the acquisition expenditure.

Tax and insurance are payable on equipment in some cases. These costs rarely vary according to transport distances or operating periods (for internal transport) and are therefore regarded as fixed costs. Operator, equipment and goods insurance exist also.

In this analysis, **costs of wages and fringe benefits** are regarded as fixed. In some cases, operators may be used for other, equally qualified tasks. Total wage costs may then be calculated as a cost per hour. Fringe benefit costs are calculated using a percentage added to the gross wage. Swedish legislation decreed an addition of approximately 31 per cent for 1980. In addition, there are contractual additions of 5.4 per cent for workers and 11.3 per cent for salaried staff.

Other fixed costs include costs of fixed auxiliary equipment such as storage premises and battery charging units. Environmental costs are difficult to determine but should also be included. They may relate to insulation against noise, ventilation or, possibly, heat losses.

Fixed administration costs may also be apportioned to equipment. They include supervision, administration and checking of operations, maintenance programmes and other equipment-related work.

Cost of energy is also a **variable operating cost.** A continuous follow-up requires that transport distances or effective operating periods are assessed relative to energy consumption.

Maintenance and repair costs obviously depend on how the equipment is utilized, i.e., they are work dependent. 'Maintenance' here means steps taken in connection with pre-arranged inspection programmes and includes some repairs as well as servicing, washing, lubricating, battery checking, etc. Work done by the operator should also be included. Repair and maintenance costs may advantageously be recorded jointly. Repairs are normally distinct from maintenance in that they are unplanned. As a consequence, a downtime cost sometimes forms part of the repair costs. This should be the case, for instance, when equivalent equipment has to be hired to cope with tasks meant for the inoperative equipment.

Maintenance and repair costs depend on many factors. Operator skill and sense of responsibility are vital, as are the environment and the utilization of productive capacity.

A **transport cost follow-up** is used as a method of controlling cost levels as well as to establish in advance future cost levels of the transport functions. The latter forms part of an operating budget for a given cost centre and is based on operating plans, maintenance plans, etc. These include forecasting values associated with operations (e.g. numbers of vehicles, operating hours, kilometres travelled, energy requirements, etc.). The operating budget can then be reconciled against actual results, through the cost follow-up. Excessive variations should then be investigated to provide information on anomalies and to prevent them occurring in the future.

A cost follow-up is facilitated if operating costs are continuously recorded, for instance, using service cards and printed forms. Information is collated for the entire cost centre and may then be used for budgeting purposes, as a cost basis for investment calculation or for controlling principally the operating costs of similar equipment.

Calculation methods

An investment generally refers to the purchase of so-called fixed capital assets, i.e., physical resources affecting a company's production capacity and having a life exceeding three years. In other words, an investment in handling equipment requires a multi-periodical decision, a fact which makes it more difficult to assess the economic consequences of an investment.

Financial investment calculation is an important part of the basis for an investment decision. When calculating the budget, the economic profitability of various investment alternatives is examined by means of studying expected cash inflow and outflow in respect of the investment. This section deals with some calculation methods and is followed by a section on formulating investment calculations, showing the calculation methods.

Investment cash flow diagram

In terms of finance an investment may be described by the cash inflow and outflow it generates over a number of years. Inflow and outflow per annum are usually collated at the end of the year and may then be described in a simple manner in a so-called investment cash flow diagram (see below). The diagram identifies cash outflow on acquisition (basic investment, B), cash inflow from the operation (I), cash outflow per annum for the operation (O), economic life expressed in years (n) and residual value (S). These components are to be found in a capital budget where the discount rate (i) is also to be found.

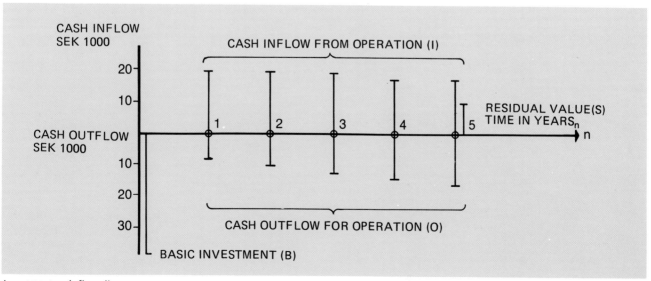

Investment cash flow diagram.

Basic investment (B) includes the cash outflow by which an investment becomes operative. To it belong procurement expenditure, training, installation and planning costs, interest on tied-up capital over the construction period and current increases in assets such as increased stockholding, pallets, etc. A basic investment need not be tied in time to the point of acquisition but may, for example, for building projects, be split into different instalments. 'Basic investment' is a budgetary concept and, therefore, requires no book-keeping equivalent.

Cash inflow and outflow (I, O) in respect of the operation are calculated per annum and are based on the operating plan, with assumptions and forecasts of production capacity, manpower requirements and wage developments, repairs and maintenance, etc. The revenue side is particularly difficult to handle. When investing in the removal of bottlenecks, or in equipment which utilizes overcapacity from a previous investment, yields may be extremely high and may thus eliminate more long-term equipment solutions. In these cases, it is assumed that equally large revenues accrue from all the alternatives, and the calculations centre on the cost aspects of the budget. Usually, the lowest annual costs or lowest total costs over the period of usage are held to be decisive.

Economic life (n) means the period over which an investment is profitable. To determine in advance in purely practical terms the economic life of new equipment is no easy task. Normally, therefore, it is done entirely on the basis of experience and based on assessment of similar equipment in the company. Most companies establish values for economic life for use in calculations. These figures are revised continuously.

The **technical life** of a piece of equipment is the upper limit of its economic life, i.e., the period during which the equipment can be used for production purposes, from a purely technical point of view. This limit is also difficult to determine as, in most cases, equipment may be repaired repeatedly without serviceability becoming impaired. This also applies to new, untested equipment where the purchaser must rely on statements made by the supplier.

At the end of its economic life, equipment will still retain some value through disposal on the second-hand market or at least as scrap. The revenue accrued is called **residual value** (S). From the residual value are subtracted the cost of dismantling, removal, etc. In view of such evolving circumstances as inflation, markets, technical advances, etc., residual value is not easy to establish.

Discount rate (i) is difficult to determine in practice. Where availability of investment capital is limited, the alternative utilization value of the capital is used as a guide. If, for instance, an alternative is to place the capital in a bank deposit account giving 12 per cent interest, the discount rate will be 12 per cent. Where the alternative is an investment yielding 15 per cent on the capital, the discount rate will be 15 per cent. It is also possible to decide the discount rate requirement by the 'price' that has to be paid for the use of the capital. This capital may be acquired internally or borrowed externally, the discount rate being adjusted to attendant costs, such as dividend on available share capital or lending rates of interest. In practice, discount rates of 10—20 per cent before tax are applicable to handling equipment.

The pay-off method

This method means that the pay-off period is calculated. In its simplest version, investment (B) is put against *average* annual yield (Y), i.e., the difference between cash inflow and outflow. The pay-off period (T) thus is the time it takes for B to be paid off.

$$T \text{ (years)} = \frac{B \text{ (SEK)}}{Y \text{ (SEK/year)}}$$

The annuity method

With the annuity method, the basic investment, possibly reduced by the present assessment of the residual value, is converted into a constant annual cost (an annuity) in accordance with the formula below. Annuity factor (a_f) values are listed in Table A below.

Table A

Annuity, i.e., the amount to be paid at the turn of each year over 1 to 50 years to amortize SEK 1.								Annuity factor $= \dfrac{i\,(1+i)^n}{(1+i)^n-1}$	
Years (n)	**Rate of interest (i)**								
	4 %	**5 %**	**6 %**	**8 %**	**10 %**	**12 %**	**15 %**	**18 %**	**20 %**
1	1.04000	1.05000	1.06000	1.08000	1.10000	1.12000	1.15000	1.18000	1.20000
2	0.53020	0.53780	0.54544	0.56077	0.57619	0.59170	0.61512	0.63872	0.65455
3	0.36035	0.36721	0.37411	0.38803	0.40211	0.41635	0.43798	0.45992	0.47473
4	0.27549	0.28201	0.28859	0.30192	0.31547	0.32923	0.35020	0.37174	0.38629
5	0.22463	0.23097	0.23740	0.25046	0.26380	0.27741	0.29832	0.31978	0.33428
6	0.19076	0.19702	0.20336	0.21632	0.22961	0.24323	0.26424	0.28591	0.30071
7	0.16661	0.17282	0.17914	0.19207	0.20541	0.21912	0.24036	0.26236	0.27742
8	0.14853	0.15472	0.16104	0.17401	0.18744	0.20130	0.22285	0.24524	0.26061
9	0.13449	0.14069	0.14702	0.16008	0.17364	0.18769	0.20957	0.23239	0.24808
10	0.12329	0.12950	0.13587	0.14903	0.16275	0.17698	0.19925	0.22251	0.23852
11	0.11415	0.12039	0.12679	0.14008	0.15396	0.16842	0.19107	0.21478	0.23110
12	0.10655	0.11283	0.11928	0.13270	0.14676	0.16148	0.18448	0.20863	0.22526
13	0.10014	0.10646	0.11296	0.12652	0.14073	0.15568	0.17911	0.20369	0.22062
14	0.09467	0.10102	0.10758	0.12130	0.13575	0.15087	0.17169	0.19968	0.21689
15	0.08994	0.09634	0.10296	0.11683	0.13147	0.14682	0.17102	0.19640	0.21388
16	0.08582	0.09227	0.09895	0.11298	0.12782	0.14339	0.16795	0.19371	0.21144
17	0.08220	0.08870	0.09544	0.10963	0.12466	0.14046	0.16537	0.19149	0.20944
18	0.07899	0.08555	0.09236	0.10670	0.12193	0.13794	0.16319	0.18964	0.20781
19	0.07614	0.08275	0.08962	0.10413	0.11955	0.13576	0.16134	0.18810	0.20646
20	0.07358	0.08024	0.08718	0.10185	0.11746	0.13388	0.15976	0.18682	0.20536
25	0.06401	0.07095	0.07823	0.09368	0.11017	0.12750	0.15470	0.18292	0.20212
30	0.05783	0.06505	0.07265	0.08883	0.10608	0.12414	0.15230	0.18126	0.20085
40	0.05052	0.05828	0.06646	0.08386	0.10226	0.12130	0.15056	0.18024	0.20014
50	0.04655	0.05478	0.06344	0.08174	0.10086	0.12042	0.15014	0.18005	0.20002

Where the annual cash inflow and outflow totals vary, they must be converted into present values and added up. The total is translated into a constant annual yield for the budget period. An investment will be profitable if the annuity does not exceed the annual constant yield (I−O), i.e., if:

$$\left(B - \frac{S}{(1+i)^n}\right)\left(\underbrace{\frac{i\,(1+i)^n}{(1+i)^n-1}}_{a_f}\right) \leqslant I - O$$

where B = basic investment (SEK)
S = residual value (SEK)
i = discount rate
n = economic life (years)
I = cash inflow from operation (SEK)
O = cash outflow for operation (SEK)
t = time.

The present value method

With the present value method all payments are discounted to the time of acquisition. Present value factors are listed in Tables B and C. The present values of cash inflow and outflow, residual value, etc., are then added up to discover whether or not the total (Co) is positive. If it is negative, the investment is unprofitable. The method is also called the capital value method after the name of the above total. Mathematically the method may be written

$$Co = \sum_{t=1}^{n} \left[(I_t - O_t)(1+i)^{-t} \right] + S(1+i)^{-n} - B \geqslant 0$$

Table B

| | Present value of SEK1 payable after n years | | | | | | | [Present value factor $= (1+i)^{-n}$] | | |

Years (n)	Rate of interest (I)									
	4 %	5 %	6 %	8 %	10 %	12 %	15 %	18 %	20 %	25 %
1	0.9615	0.9524	0.9434	0.9259	0.9091	0.8929	0.8696	0.8475	0.8333	0.8000
2	0.9346	0.9070	0.8900	0.8673	0.8264	0.7972	0.7561	0.7182	0.6944	0.6400
3	0.8890	0.8668	0.8396	0.7938	0.7513	0.7118	0.6575	0.6036	0.5787	0.5120
4	0.8548	0.8227	0.7921	0.7350	0.6830	0.6355	0.5718	0.5158	0.4823	0.4096
5	0.8219	0.7835	0.7473	0.6806	0.6209	0.5674	0.4972	0.4371	0.4019	0.3277
6	0.7903	0.7462	0.7050	0.6302	0.5645	0.5066	0.4323	0.3704	0.3349	0.2621
7	0.7599	0.7107	0.6651	0.5835	0.5132	0.4523	0.3759	0.3139	0.2791	0.2097
8	0.7307	0.6768	0.6274	0.5403	0.4665	0.4039	0.3269	0.2660	0.2326	0.1678
9	0.7026	0.6446	0.5919	0.5002	0.4241	0.3606	0.2843	0.2255	0.1938	0.1342
10	0.6756	0.6139	0.5584	0.4632	0.3855	0.3220	0.2472	0.1911	0.1615	0.1074
11	0.6496	0.5847	0.5268	0.4289	0.3505	0.2875	0.2149	0.1619	0.1346	0.0859
12	0.6246	0.5568	0.4970	0.3971	0.3186	0.2567	0.1869	0.1372	0.1122	0.0687
13	0.6006	0.5303	0.4688	0.3677	0.2897	0.2292	0.1625	0.1163	0.0935	0.0550
14	0.5775	0.5051	0.4423	0.3405	0.2633	0.2046	0.1413	0.0985	0.0779	0.0440
15	0.5553	0.4810	0.4173	0.3152	0.2394	0.1827	0.1229	0.0835	0.0649	0.0352
16	0.5339	0.4581	0.3936	0.2919	0.2176	0.1631	0.1069	0.0708	0.0541	0.0281
17	0.5134	0.4363	0.3714	0.2703	0.1978	0.1456	0.0929	0.0600	0.0451	0.0225
18	0.4936	0.4155	0.3503	0.2502	0.1799	0.1300	0.0808	0.0508	0.0376	0.0180
19	0.4746	0.3957	0.3305	0.2317	0.1635	0.1161	0.0703	0.0431	0.0313	0.0144
20	0.4564	0.3769	0.3118	0.2145	0.1486	0.1037	0.0611	0.0365	0.0261	0.0115
25	0.3751	0.2953	0.2330	0.1460	0.0923	0.0588	0.0304	0.0160	0.0105	0.0038
30	0.3083	0.2314	0.1741	0.0994	0.0573	0.0334	0.0151	0.0070	0.0042	0.0012
40	0.2083	0.1420	0.0972	0.0460	0.0221	0.0107	0.0037	0.0013	0.0007	0.0001
50	0.1407	0.0972	0.0543	0.0213	0.0085	0.0035	0.0009	0.0003	0.0001	0.0000

Table C

| Total present value of SEK 1 payable at the turn of the year over the following 1 to 50 years | | | | | | | Present value factor $= \dfrac{(1+i)^n - 1}{i(1+i)^n}$ | | | |

Years (n)	Rate of interest (i)									
	4 %	5 %	6 %	8 %	10 %	12 %	15 %	18 %	20 %	25 %
1	0.962	0.952	0.943	0.926	0.909	0.893	0.870	0.847	0.833	0.800
2	1.886	1.859	1.833	1.783	1.736	1.690	1.626	1.566	1.528	1.440
3	2.775	2.723	2.673	2.577	2.487	2.402	2.283	2.174	2.107	1.952
4	3.630	3.546	3.465	3.312	3.170	3.037	2.855	2.690	2.589	2.362
5	4.452	4.329	4.212	3.993	3.791	3.605	3.352	3.127	2.991	2.689
6	5.242	5.076	4.917	4.623	4.355	4.111	3.785	3.498	3.326	2.951
7	6.002	5.786	5.582	5.206	4.868	4.564	4.160	3.812	3.605	3.161
8	6.733	6.463	6.210	5.747	5.335	4.968	4.487	4.078	3.837	3.329
9	7.435	7.108	6.802	6.247	5.759	5.328	4.772	4.303	4.031	3.463
10	8.111	7.722	7.360	6.710	6.145	5.650	5.019	4.494	4.193	3.570
11	8.760	8.306	7.887	7.139	6.495	5.938	5.234	4.656	4.327	3.656
12	9.385	8.863	8.384	7.536	6.814	6.194	5.421	4.793	4.439	3.725
13	9.986	9.394	8.553	7.904	7.103	6.424	5.583	4.910	4.533	3.780
14	10.563	9.899	9.295	8.244	7.367	6.628	5.725	5.008	4.611	3.824
15	11.118	10.380	9.712	8.559	7.606	6.811	5.847	5.092	4.676	3.859
16	11.652	10.838	10.106	8.851	7.824	6.974	5.954	5.162	4.730	3.887
17	12.166	11.274	10.477	9.122	8.022	7.120	6.047	5.222	4.775	3.910
18	12.659	11.690	10.828	9.372	8.201	7.250	6.128	5.273	4.812	3.928
19	13.134	12.085	11.158	9.604	8.365	7.366	6.198	5.316	4.844	3.942
20	13.590	12.462	11.470	9.818	8.514	7.469	6.259	5.353	4.870	3.954
25	15.622	14.094	12.783	10.679	9.077	7.643	6.464	5.467	4.948	3.985
30	17.292	15.372	13.765	11.258	9.427	8.055	6.566	5.517	4.979	3.995
40	19.793	17.159	15.046	11.925	9.779	8.244	6.642	5.548	4.997	3.999
50	21.482	18.256	15.762	12.233	9.915	8.304	6.661	5.554	4.999	4.000

Internal rate of return

The internal rate of return (r) on an investment is the rate at which the total of all cash inflow and outflow discounted to the time of acquisition (= the capital value, see above) is zero. If the residual value can be disregarded and the annual inflow and outflow are constant, this may be written as

$$B - (O-I) \sum_{t=1}^{n} (1+r)^{-t} = 0$$

The internal rate of return becomes a measure of the percentage yield of the investment. The investment is profitable, therefore, if the internal rate of return is higher than the set discount rate of interest for the investment. As can be seen, it is necessary to proceed by trial and error, using a table of present value totals and extrapolation in order to determine the internal rate of return These calculations may sometimes be rather cumbersome, so the work has been simplified by the production of prepared nomograms. The nomogram below may be used for quickly examining the internal rate of return on an investment, the following assumptions obtaining. The basic investment (B) is made at the beginning of year 0, the differential cash inflow (the average annual gross profit (a)) is constant each year, and the residual value is put at zero. The pay-off period (B/a) is calculated, and the internal rate of return is then read as indicated for any given or assumed economic life.

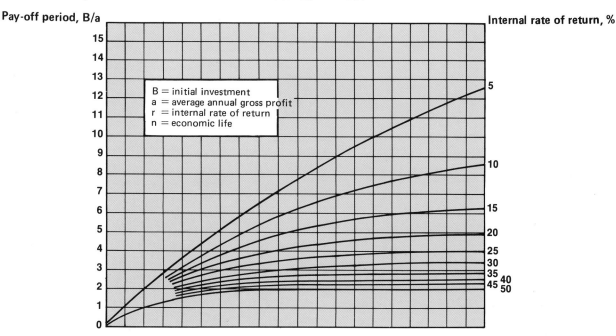

Graph for determining internal rates of return.

Formulating investment calculations for various investment alternatives

Assessing alternative new investments

For a **new investment,** choose in principle the alternative with the highest profitability ratio for the various methods used. When comparing alternatives, differences in economic life and in the amounts of basic investment should be observed.

When comparing alternatives of **unequal basic investments,** examine at what interest rate the difference can be placed. If it is possible to invest the difference in accordance with the set discount rate of interest, the comparison will not be affected. If not, the more expensive alternative will be burdened with the internal rate of return on the differing amount or, with the present value and annuity methods, with the differing interest charge.

For different economic lives of the investments a similar rationale applies to the present value method. It is necessary here to explain how the capital could be utilized over the time difference. Differences in the economic life of the investments

mean that the annuity method cannot be used for a comparison. The annuity must be calculated on the same basis (= economic life) in order to be comparable. However, where a continual repetition of this investment is assumed, i.e., continual replacement investments, it is feasible to use the method. Where the economic life is the same in each case, all the methods can be used.

With both the present value and the annuity method, profitability is expressed in absolute figures. For differing investment amounts these profitability ratios are sometimes expressed in relation to the basic investment.

The **present value ratio** (capital value ratio) is the ratio between present value and basic investment. The alternative having the highest ratio is regarded as the more advantageous.

The **annuity ratio** is calculated as the difference between the yield and annuity of the basic investment expressed in relation to the basic investment. The ratio is regarded as being a measure of excess interest, i.e., the contribution in excess of the discount interest generated by the alternative. Even in this case the higher ratio is the more advantageous.

Replacement investment calculations

Considering the replacement of existing equipment is not only a question of establishing whether or not new equipment will prove more profitable but also when replacement should be effected in order to be as profitable as possible. Where the service life of old equipment is at an end, the same methods as for new investments may be used.

Where the object to be replaced has still some service life, there will be problems of determining the basic investment. As investment calculations involve measuring changes in the cash flow, the basic investment must equal the cash inflow obtained when the equipment is sold or scrapped. The book value cannot be used. In other respects, the residual economic life and other calculation components are used in the same way.

Handling calculation uncertainties

Calculation results are no better than the input data used for calculating. As supporting data are often based on forecasts and assumptions about the future, there will obviously exist some uncertainty. Various methods are used to reduce this uncertainty.

When calculating the outcome **parallel calculations** may be formulated. Using a standard case as the starting point, the best and the least satisfactory alternatives are then calculated to get some idea of the extent of variation in the result.

A better way is varying one calculation component at a time, e.g., economic life, to see how this affects the final result. This method measures the sensitivity of the calculated result to defects in the components (sensitivity analysis).

Adding margins to the various components 'in order to be on the safe side' is not recommended. Often, only the person making the calculations understands these intricacies. Thus, any comparison between alternatives will be faulty, and more profitable investments may be eliminated in the final assessment.

Choice of calculation method

The scope of the calculations and the extent of user knowledge are of great importance when choosing the calculation method. The ability of staff to master a calculation method is more important than the method's sophistication. This is inevitable in view of the large number of small investments made which require uniform assessment. On the other hand, the additional work involved in the calculations carries less weight in comparison with the work put into producing supporting data. The choice of calculation method also depends on the basic decision data required by the individual decision makers in the company.

The great advantage of the **pay-off method** is that it is simple to understand and use and is therefore frequently used for small investments or for initial approximations. The pay-off period is also the best measure of potential liquidity problems caused by the investment. Its disadvantage is that it does not take into account interest on available capital and economic life subsequent to the pay-off period. Economic life of different duration, therefore, cannot be assessed in a comparison where the pay-off period is the same for both alternatives. Besides, no account is taken of cash inflow after the pay-off period has ended.

Unlike the pay-off method the **internal rate of return method** is fairly complicated and laborious. The internal rate of return, however, is seen as the variable to be maximized. In theoretical terms, the internal rate of return does not describe the cost of allocated capital, as does budget rate of interest. However, this information is unnecessary for ranking alternatives, and therefore the lack of it is no disadvantage. The method, however, depends from the calculation point of view on the discount rate of interest figure being correct.

The **present value method** is advantageous for comparisons between objects of different economic life. The profitability is expressed as an absolute figure. This method is to be preferred to the internal rate of return method when individual investments are compared with the zero alternative (= no investment).

The **annuity method** pre-supposes comparatively constant cash flow which has no equivalent in practice. Not being able to use this method in cases of varied economic life, without assuming a perpetual repetition of the investment, is obviously unsatisfactory. As with the present value method, the profitability ratio is an absolute figure, a fact which makes comparison difficult. The direct dependence of the annual costs on economic life and on determining the residual value makes the calculated result sensitive to rapid external variations.

Forms of financing

Industry today is characterized by a reduced level of self-financing and the weeding out of activities extraneous to an operation; as a result, new forms of investment financing have emerged. The most common forms are self-financing (paid-up capital), supplier's credit (instalment payment), bank credit (external borrowing) and, in recent years, various forms of long-term hire.

Self-financing requires the company to generate a comparatively large and even operating surplus. This frequently causes difficulties, especially for expanding companies, whose capital resources are used for a multitude of urgent purposes, and whose expansion would, thereby, be limited.

Supplier's credit often results in a bargaining position less advantageous than one based on cash payment, and this may affect the purchase price level. In addition, interest charges will be high in many cases.

Bank credit, when available, is often a low cost form of financing. Securities acceptable to the bank must be guaranteed. Where the company is deficient in securities any of the credit forms listed below may be used:

Leasing — The purchaser and the supplier negotiate a price and the terms of delivery. Subsequently, a finance company pays cash to the supplier and takes over ownership of the product after which the purchaser pays charges to the finance company for using the product. In this way the ownership and the purchaser's credit-worthiness become 'securities' acceptable to the finance company.

Supplier's leasing — Similar to the above method except that there is no finance company as middleman. The supplier finances his own product and retains ownership as security. The purchaser is invoiced monthly or quarterly.

Contract hire rental — In addition to the financing organized in the same way as a leasing agreement, the supplier undertakes to handle service, repairs, reconditioning, spares, stand-by machinery during stoppages, etc. The customer rents functions rather than products, and the supplier must ensure that the production capacity of the equipment is available for the entire contract period.

Some suppliers also provide **short time hire** intended to meet customer's more ad hoc needs, e.g., during stocktaking, to compensate for equipment downtime, at seasonal peaks, etc.

In addition to supplying the equipment, the undertakings normally include the following items:

Some training		
Freight		
Standby machinery during stoppages		
Maintenance adminstration		Rental
Spares		
Repairs		
Service		
Insurance		Short-term
Interest	Leasing	hire
Depreciation		
Tyre costs, if any		
Battery costs, if any		

In terms of tax and liquidity the above financing methods obviously create dissimilar investment cash flow diagrams. One reason is that rentals and loan interest may be deducted for tax purposes as business expenditure. Liquidity depends on a daily balance between cash inflow and outflow, and a firm may end up in a situation where liquidity is more important than investment profitability. Spreading the cost as, for instance, with a monthly rental may prove more advantageous than a large non-recurring cost.

The basic idea behind leasing and rental is that it is the use of the equipment, not its ownership, which is of the essence. In principle, most types of fixed assets could be leased. However, this does not apply to real estate. Rental is based on the acquisition cost of the equipment, and the monthly cost is normally indicated as a percentage of the former. In principle, the duration of the leasing agreement must not exceed the economic life of the equipment and normally varies between 3 and 5 years. The agreement having expired, the customer is entitled to extend it at a revised annual rental.

From a calculation point of view, leasing becomes more advantageous the higher the discount rate. For most firms, rental also has benefits which accrue from unloading problems extraneous to the operation.

A contract hire agreement offers security against stoppages and downtime, backed by supplier service. It also simplifies the lessee's administrative work.

Calculation examples

Assume that a company is thinking of changing from manual handling to using three trucks. The economies consist mainly of reducing five manual workers to three truck drivers. The trucks have diesel engines and, with single shift working, their operating time per annum is estimated at 1,200 hours. The following costs, it is calculated, will accrue:

Basic investment (B)

	SEK
Purchase price after discount	300,000
(3 trucks @ SEK 100,000)	
Training	6,000
(3 drivers @ SEK 2,000)	
Freight (3 @ SEK 1,000)	3,000
Installing ventilation	7,000
	B = 316,000

Residual value

10% of purchase price	30,000

Discount interest (i), SEK/year	20% before tax
Economic life (n)	8 years
Cash inflow (I)	SEK/year
5 man years @ SEK 90,000 dispensed with (total wage cost)	450,000

Cash outflow (O), SEK/year
Operating costs

Fixed:	3 drivers @ SEK 90,000	270,000
	insurance	3,000
	ventilation and heating costs	6,000
Variable:	(1,200 hours/year)	
	fuel	3,600
	replacement engine	5,400
	maintenance and repairs	
	including cost of downtime	54,000
	tyres	9,000
		350,000

The pay-off method

The pay-off period (T) is calculated according to $T = B/Y$. The yield (Y) equals the annual cash inflow less the cash outflow $(I - O)$. If constant conditions are assumed, it follows that

$$T = \frac{B}{Y} = \frac{31.6}{45 - 35} = 3.2 \text{ years}$$

When the pay-off period is so long, calculating it by the pay-off method is somewhat unreliable as a basis for decision-making. However, constant conditions over such a period are assumed to be

The annuity method
This method is based on producing equal annual costs for the calculation period and comparing them against an equally constant annual revenue. As stated, the profitability is indicated by the formula below:

$$B - \frac{S}{(1+i)^n} \, a_f \, \frac{20\%}{8 \text{ years}} \leqslant 1 - 0$$

The annuity factor (a_f) is found in Table A (page 33) and represents a 20 per cent discount rate. The eight-year economic life equals 0.26061. The present value of the residual value at zero years is obtained with the aid of Table B and is

$$S \times 0.2326 = \text{SEK } 3,489$$
The annuity $(B - 6978) \, 0.26061 = \text{SEK } 81,443$

This is less than $(I - O) = \text{SEK } 99,000$, indicating that the investment is profitable.

A cost per hour calculation for one of the above diesel trucks can be made using the annuity method. The total number of driving hours is assumed to be 1,200 hours/year.

		SEK per truck hour
Fixed	Basic investment less residual value at zero years 105,333 − 2,326 = 103,000 8 year life, 20 per cent interest Constant value of depreciation + interest = 103,000 × 0.2606 (annuity) ≈ 26,800 Hourly cost for depreciation and interest (1,200 hours/year)	22
	Ventilation and heating (soundproofing, if any)	5
Variable	Repairs and maintenance (labour and spare parts)	15
	Power (including standby engine, or battery, if any)	2.5
	Tyres	2.5
		47

Total hourly cost
Add total driver's wage
SEK 90,000/1,200 — 75

The present value method
All cash inflow and outflow pertaining to the investment are converted to year zero values and added together. Where the total is larger than zero, profitability is indicated.

$$\sum_{t=1}^{n} (I_t - O_t)(1+i)^{-t} + S(1+i)^{-n} - B \geqslant 0$$

An amount constant over a period has a total which may be expressed as its present value at zero years. Using Table C and at 20 per cent discount rate over eight years

$(I - O) \, 3.837 + S \times 0.2326 - B =$
$380,000 + 3,490 - 316,000 = 67,500$

The total is larger than zero, i.e., the investment is profitable.

The internal rate of return method
The internal rate of return is determined by means of the graph on page 35. B/Y was earlier calculated at approximately 3.2 years. With an economic life of eight years, the graph shows the internal rate of return to be approximately 27 per cent. To achieve profitability it is necessary for this rate to exceed the 20 per cent discount rate. This the internal rate of return does.

Tax effects of different forms of financing

In the example described above, the company could choose between financing the acquisition of the trucks with its own capital, an external bank loan or a contract hire agreement. A preference for contract hire rather than a leasing arrangement can be justified in that the company had no experience of operating and maintaining trucks. Hire-purchase was also considered but could be avoided on the strength of a positive response from the bank. Both these types of financing offered repayment over three years while the supplier's effective annual rate of interest was 20 per cent as against the bank's flat-rate loan interest of 15 per cent.

Table D

Years	Final value of SEK 1 after 1 to 50 years. [Interest factor $(1+i)^n$]								
(n)	Interest rate (i)								
	4 %	5 %	6 %	8 %	10 %	12 %	15 %	18 %	20 %
1	1.040	1.050	1.060	1.080	1.100	1.120	1.150	1.180	1.200
2	1.082	1.103	1.124	1.166	1.210	1.254	1.323	1.392	1.440
3	1.125	1.158	1.191	1.260	1.331	1.405	1.521	1.643	1.728
4	1.170	1.216	1.262	1.360	1.464	1.574	1.749	1.939	2.074
5	1.217	1.276	1.338	1.469	1.611	1.762	2.011	2.288	2.488
6	1.265	1.340	1.419	1.587	1.772	1.974	2.313	2.700	2.986
7	1.316	1.407	1.504	1.714	1.949	2.211	2.660	3.185	3.583
8	1.369	1.477	1.594	1.851	2.144	2.476	3.059	3.759	4.300
9	1.423	1.551	1.689	1.999	2.358	2.773	3.518	4.435	5.160
10	1.480	1.629	1.791	2.159	2.594	3.106	4.046	5.234	6.192
11	1.539	1.710	1.898	2.332	2.853	3.479	4.652	6.176	7.430
12	1.601	1.796	2.012	2.518	3.138	3.896	5.350	7.288	8.916
13	1.663	1.886	2.133	2.720	3.452	4.363	6.153	8.599	10.699
14	1.732	1.980	2.261	2.937	3.797	4.887	7.076	10.147	12.839
15	1.801	2.079	2.397	3.172	4.177	5.474	8.137	11.974	15.407
16	1.873	2.183	2.540	3.425	4.595	6.130	9.358	14.129	18.488
17	1.948	2.292	2.693	3.700	5.054	6.856	10.761	16.672	22.186
18	2.026	2.407	2.854	3.996	5.560	7.690	12.375	19.673	26.623
19	2.107	2.527	3.026	4.316	6.116	8.613	14.232	23.214	31.948
20	2.191	2.653	3.207	4.661	6.727	9.646	16.367	27.393	38.338
25	2.666	3.386	4.292	6.848	10.835	17.000	32.919	62.669	95.396
30	3.243	4.322	5.743	10.063	17.449	29.960	66.212	143.371	237.376
40	4.801	7.040	10.286	21.725	45.259	93.051	267.864	750.379	1469.772
50	7.107	11.467	18.420	46.902	117.391	289.002	1083.657	3927.357	9100.438

British Standards

Inevitably, materials handling equipment and practices feature prominently in the standards issued by the British Standards Institution, Linford Wood, Milton Keynes MK14 6LE. The following selection concentrates on operational aspects rather than design criteria.

CRANES

BS 327:1964
Power-driven derrick cranes Gr 7

BS 357:1958
Power-driven travelling jib cranes (rail-mounted low carriage type) Gr 7

BS 466:1960
Electric overhead travelling cranes for general use in factories, workshops and warehouses Gr 7

BS 1757:1981
Power-driven mobile cranes Gr 8

BS 2452:1954
High pedestal or portal jib cranes Gr 7

BS 2799:1974
Power-driven tower cranes for building and engineering construction Gr 8

BS 3579:1963
Heavy duty electric overhead- travelling and special cranes for use in steel works Gr 8

BS 3810:——
Glossary of terms used in materials handling

BS 3810:Part 4:1968
Terms used in connection with cranes Gr 5

CHAIN AND TERMINAL ATTACHMENTS

BS 2903:1980   ± ISO 2141, 2766
Higher tensile steel hooks for chains, slings, blocks and general engineering purposes Gr 7

BS 4278:1968
Eyebolts for lifting purposes Gr 5

BS 4283:1968
Swivels for lifting purposes Gr 5

BS 4291:1968
Treble-swivel cargo hooks for ships' union purchase Gr 4

BS 4654:1970   ± ISO 2308
Hooks for lifting freight containers of up to 30 tonnes Gr 3

BS 4831:1972   ≠ ISO/R 1082
Shackle type connector units for high tensile steel chain for mining type conveyors Gr 4

BS 4942:——
Short link chain for lifting purposes

BS 5237:1975
Lifting twistlocks Gr 6

CONVEYORS

BS 490:——
Conveyor and elevator belting

BS 490:Part 10:Section 10.2:1983
Method for determination of full thickness tensile strength and elongation of rubber and plastics conveyor belting of textile construction Gr 2

BS 490:Part 10:Section 10.4:1983
Method for determination of adhesion strength of rubber and plastics belting of textile construction Gr 2

BS 2567:1972
Steel non-powered roller conveyors Gr 4

BS 2890:1973   ≠ ISO 1535-7, 1816, 1819, 2109
Troughed belt conveyors Gr 8

BS 3810:Part 2:1965   ≠ ISO 2148
Terms used in connection with conveyors and elevators (excluding pneumatic and hydraulic handling) Gr 5

BS 4409:——   ≠ ISO 1050, 1819, 3264
Screw conveyors

BS 4409: Part 1: 1969
Trough type for industrial use Gr 4

BS 4409:Part 2:1969
Portable and mobile tubular type (augers) for agricultural and light industrial use Gr 3

BS 4531:1969   ≠ ISO 1819, 2387-8
Portable and mobile troughed belt conveyors Gr 4

BS 5667:——
Continuous mechanical handling equipment — Safety requirements

BS 5667:Part 2:1979   = ISO 5028
Loose bulk materials: pneumatic handling installations Gr 2

BS 5667:Part 3:1979   = ISO 5029
Loose bulk materials: storage equipment fed by a pneumatic handling system Gr 2

BS 5667:Part 4:1979   = ISO 5030
Loose bulk materials: mobile suction pipes suspended from derrick jibs used in pneumatic handling Gr 2

BS 5667:Part 6:1979   = ISO 5032
Loose bulk materials: rotary feeders used in pneumatic handling Gr 2

BS 5667:Part 7:1979   = ISO 5033
Loose bulk materials: rotary drum feeders and rotary vane feeders Gr 2

BS 5667:Part 10:1979   = ISO 5036
Loose bulk materials: vertical screw conveyors Gr 2

BS 5667:Part 11:1979   = ISO 5037
Unit loads: fixed slat conveyors (metal or wood) with horizontal shafts Gr 2

BS 5667:Part 12:1979   = ISO 5038
Unit loads: mobile slat conveyors (metal or wood) with horizontal shafts Gr 2

BS 5667:Part 13:1979   = ISO 5039
Unit loads: arm elevators and push bar conveyors Gr 2

BS 5667:Part 14:1979   = ISO 5040
Unit loads: live roller conveyors (with positive or friction drive) Gr 2

BS 5667:Part 15:1979   = ISO 5041
Unit loads: crate-carrying chain conveyors having biplanar chains for flat-bottomed unit loads Gr 2

BS 5667:Part16:1979   = ISO 5042
Unit loads: slat band chain conveyors Gr 2

BS 5667:Part 17:1979   = ISO 5043
Unit loads: continuous plate conveyors (horizontal) Gr 2

BS 5767:1979   = ISO 5285
Guide for storage and handling of conveyor belts Gr 2

BS 5934:1980   = ISO 5048
Method for calculation of operating power and tensile forces in belt conveyors with carrying idlers on continuous mechanical handling equipment Gr 7

BS 6318:——
Bucket elevators

BS 6318:Part 1:1982   = ISO 7190
Classification of bucket elevators Gr 4

BS 6318:Part 2:1982   = ISO 5050
Dimensions of vertical bucket elevators with calibrated round steel link chains Gr 2

BS 6318:Part 3:1982   = ISO 5051
Dimensions of deep elevator bucket with flat rear wall Gr 2

INDUSTRIAL TRUCKS

BS 3726:1978   = ISO 1074
Specification for counterbalanced lift trucks — stability — basic tests Gr 3

BS 4155:1967
Dimensions of pallet trucks Gr 2

BS 4337:1968   ≠ ISO/R 642;
± ISO 938, 1756
Principal dimensions of hand-operated stillage trucks Gr 2

BS 4338:1968   ≠ ISO/R 1214
Rated capacities of fork lift trucks Gr 2

BS 4339:1968   ≠ ISO 1084
Rating of industrial tractors Gr 2

BS 4436:1978   = ISO 3184
Specification for reach and straddle fork lift trucks — stability tests Gr 5

BS 5639:——
Fork arms for fork lift trucks

BS 5639:Part 1:1978   = ISO 2331
Vocabulary for hook-on type fork arms Gr 3

BS 5639:Part 2:1978   = ISO 2328
Mounting dimensions of hook-on type fork arms and fork carriers Gr 3

BS 5639:Part 3:1978
Recommendations for dimensions of fork arms Gr 2

BS 5639:Part 5:1978
Guide for inspection and repair of fork arms in service Gr 2

BS 5933:1980   = ISO 6055
Overhead gusrds for high-lift rider trucks Gr 2

SLINGS

BS 3458:1962
Alloy steel chain slings Gr 7

BS 3481:——
Flat lifting slings

BS 3481:Part 1:1962
Wire coil flat slings Gr 6

BS 3481:Part 3:1974
Disposable flat lifting slings Gr 6

BS 6166:1981
Recommendations for rating of lifting gear for general purposes Gr 3

BS 6210:1983
Code of practice for the safe use of wire rope slings for general lifting purposes Gr 6

PULLEY BLOCKS

BS 1692:1971
Gin blocks Gr 3

BS 3243:1973
Hand-operated chain pulley blocks Gr 6

BS 3701:1964
Hand-operated plate-sided winches Gr 4

BS 3810:Part 6:1973
Terms used in connection with pulley blocks Gr 7

BS 4018:1966
Pulley blocks for use with wire rope for a maximum lift of 25 tons/in² combination Gr 4

BS 4344:1968
Pulley blocks for use with natural and synthetic fibre ropes Gr 4

BS 4536:1970
Heavy duty pulley blocks for use with wire ropes Gr 4

MA 47:1977
Code of practice for ships' cargo blocks Gr 5

MA 48:1976
Code of practice for design and operation of ships' derrick rigs Gr 7

LIFTS, HOISTS AND ESCALATORS

BS 2655:——
Lifts, escalators, passenger conveyors and paternosters

BS 2655:Part 1:1970
General requirements for electric, hydraulic and hand-powered lifts Gr 7

BS 2655:Part 4:1969
General requirements for escalators and passenger conveyors Gr 5

BS 2655:Part 5:1970
General requirements for paternosters Gr 4

BS 2655:Part 8:1971
Modernization or reconstruction of lifts, escalators and paternosters Gr 5

BS 3810:Part 5:1971
Terms used in connection with lifting tackle Gr 6

BS 3810:Part 7:1973
Terms used in connection with aerial ropeways and cableways Gr 7

BS 4898:1973 (1980)
Chain lever hoists Gr 5

BS 5323:1980
Code of practice for scissor lifts Gr 5

BS 5655:——
Lifts and service lifts

BS 5655:Part 2:1983
Specification for hydraulic lifts Gr 3

BS 5656:1983   EN 115
Safety rules for the construction and installation of escalators and passenger conveyors Gr 8 MEE/49

BS 6109:1981
Code of practice for tail lifts Gr 6

BS 6289:——
Work platforms

BS 6289:Part 1:1982
Code of practice for mobile scissor operated work platforms Gr 7

CP 407:1972
Electric, hydraulic and handpowered lifts Gr 8

STORAGE AND TRANSPORT EQUIPMENT

BS 826:1978
Steel single tier bolted shelving (angle upright type) Gr 7

BS 2629:——
Pallets for materials handling for through transit

BS 2629:Part 2:1970
Pallets for use in freight containers Gr 4

BS 3810:Part 1:1964
Terms used in connection with pallets, stillages, hand and powered trucks Gr 6

BS 3951:1977
Freight containers

BS 3951:Part 1:Section 1.1:1979   = ISO 668
Series 1 freight containers — Classification, external dimensions and ratings Gr 3

BS 3951:Part 1:Section 1.2:1977   = ISO 1161
Specification for corner fittings for series 1 freight containers Gr 6

BS 3951:Part 1:Section 1.3:1979   = ISO 1894
Specification for minimum internal dimensions for general purpose series 1 freight containers Gr 2

BS 3951:Part 1:Section 1.4:1983   = ISO 830
Glossary of terminology Gr 6

BS 3951:Part 1:Section 1.5:1977   = ISO 3874
Guide to the handling and securing of series 1 freight containers Gr 6

BS 3951:Part 1:Section 1.6:1983   = ISO 6346
Specification for coding, identification and marking Gr 9

BS 3951:Part 2:Section 2.1:1978   = ISO 1496/1
General cargo containers Gr 7

BS 3951:Part 2:Section 2.2:1978   = ISO 1496/2
Thermal containers Gr 8

BS 3951:Part 2:Section 2.8:1978   =ISO 1496/6c
Platform based containers open-sided with complete superstructure Gr 7

BS 4345:1968
Slotted angles Gr 3

An important reference source for books on materials handling and distribution in the UK is the National Materials Handling Centre, Cranfield, Bedford MK43 0AL.

Equipment

PART

2

Materials handling is the art of building up and maintaining a physical materials flow (raw materials, semi-finished and finished products) on premises used for production, storage or distribution in the widest sense.

Materials handling may also be said to be a method of solving problems. These problems break down into the following commonly encountered sub-problems or sub-functions:

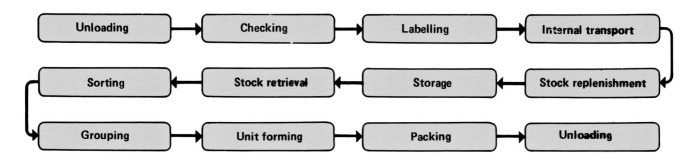

Most sub-functions are based on one or several of the four primary functions:

- Transport
- Handling
- Storage
- Unit formation

The various types of equipment presented in this part of the manual all have one thing in common: they should satisfy one or several of these primary functions.

The tabulation below categorizes the manual's contents on the basis of the primary functions:

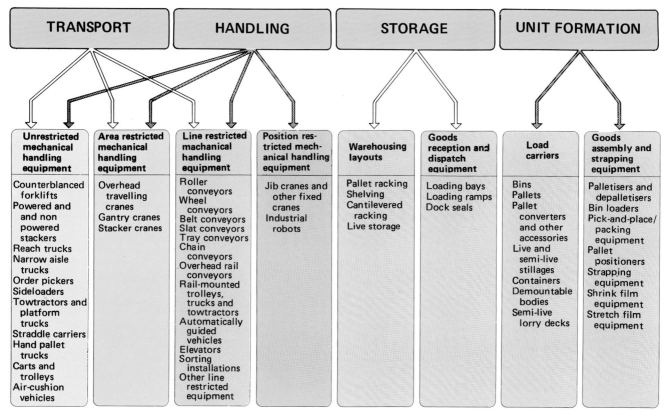

In addition to the primary classifications above, more detailed classifications appear in various parts of this manual, for instance, in respect of trucks (see 'General information on trucks') and line restricted equipment (see 'General information on line restricted materials handling equipment'). The above tabulation also includes equipment involved in the purely physical aspects of the materials flow. Equipment for labelling, control, weighing, etc., is excluded.

General information on trucks

Trucks are used to transport materials over short distances. Most trucks are also able to lift and stack materials.

Swedish Standard SMS 2795 employs the following definitions:

Powered truck A powered mechanical handling vehicle for load carrying or traction purposes.

Hand pallet truck A non-powered mechanical handling vehicle able to lift and carry loads.

Trucks have no route restrictions and are suitable for application where flexibility is important.

Where materials flow is continuous and/or predictable, line restricted devices, e.g. roller conveyors, may have advantages. However, such equipment usually requires fixed installations which are difficult to change and which hinder or prevent other materials flow. Trucks are not only often more flexible; in certain cases they are the only practical solution, e.g., for loading and unloading open lorries, containers and rail wagons.

Powered truck classification

As yet there is no generally accepted classification of powered trucks. The classification below is an attempt to impose order on the different types. A draft materials transport terminology was also published by TFK in 1981.

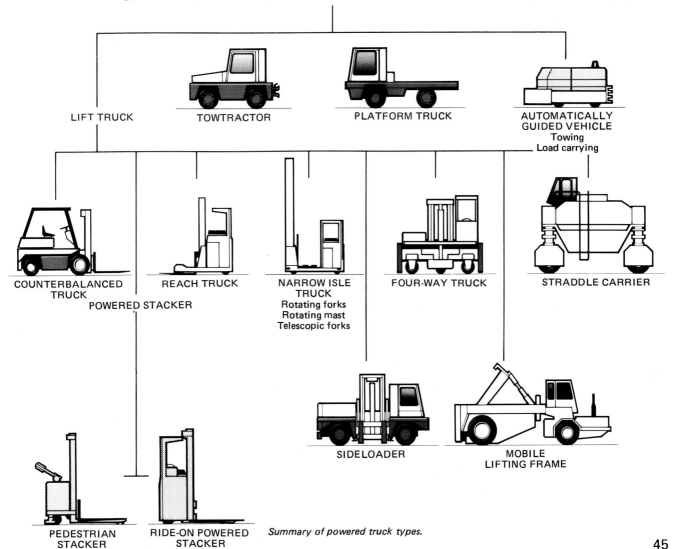

Summary of powered truck types.

45

Technical characteristics

This section describes components which are common to all types of truck. It can be used as the basis for reading subsequent chapters on the different trucks.

Chassis

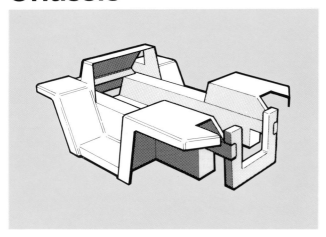

Truck chassis

The chassis is a self-supporting welded plate and beam structure. It is essential that the chassis is stable and that it is designed to give easy access for servicing and to vital components.

Power sources

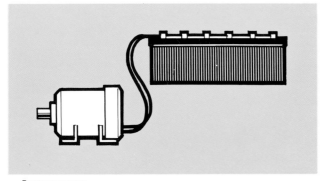

Battery power

Battery power has great environmental advantages. No exhaust fumes are emitted, noise levels are low, and vibration is slight. Consequently, battery power is suitable for indoor operations. Service costs are often slightly lower than for other power sources.

Battery storage of electric power is expensive, bulky and heavy. A battery truck is both dearer and heavier than an internal combustion engine truck. Its endurance is also inferior.

Battery capacities are generally adapted to satisfy the requirements of normal operation during an eight hour shift. For multi-shift operation, two separate battery sets are used. Breaks for partial recharging of batteries are not recommended as battery life is directly affected by the number of charging cycles. Nor is high-speed charging good for batteries.

The battery rate of charge is usually thyristor or resistor controlled. With thyristor control, the electric current is divided electronically into short pulses, the duration of which may be varied. The efficiency of such a system is high. Resistor control, by rheostat, causes comparatively large heat losses and is less efficient than thyristor control. Today (1981) thyristor control is slightly more expensive than resistor control.

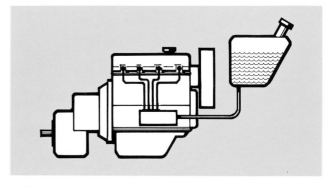

Diesel engine

A **diesel engine** truck is very powerful in relation to its weight and can therefore use many types of attachment. As it can carry large quantities of energy, operating periods between refuelling are long. Operating costs are low in comparison with other systems due to low energy consumption and low energy costs. This power source has also proved itself in other vehicle types.

Petrol engines provide similar operating advantages to those of diesels but operating costs are somewhat higher. Today (1981) petrol driven trucks are comparatively rare.

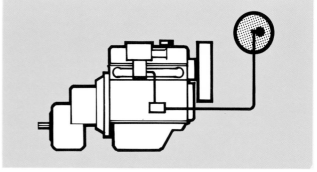

LP gas engine

On trucks powered by **LP gas,** the gas is stored under pressure in special cylinders. Refuelling, by changing a cylinder, is a quick operation. One advantage of LP gas is that it has better cold start characteristics than those of a diesel engine.

LP gas and diesel engines are similar with respect to ventilation requirements when used indoors. However, the composition of exhaust fumes is different. In diesel exhausts, sooty smoke and nitrogenous gases are the most troublesome substances emitted. Nitrogenous gases, often written as NOx, are produced by oxidization of nitrogen in atmosphere. At higher temperatures nitrogen becomes oxidized at a higher rate. As a diesel engine operates at high combustion temperatures, it produces more nitrogenous gases than other internal combustion engines. Nitrogenous gases cause no direct acute symptoms in man. In the long term they may cause bronchitis and other pulmonary illnesses.

Petrol engines produce primarily carbon monoxide, a lethal gas which initially induces symptoms of tiredness, nausea and headache.

Acceptable limits for the most common substances contained in exhaust fumes have been determined by the Swedish Industrial Safety Board. These limits aim to prevent health hazards in those exposed to air contaminated by these substances. To conform to the set limits, the premises must have satisfactory ventilation in cases where an IC engine truck is used indoors — particularly in confined spaces.

The characteristics of the various power sources are summarized in the adjacent table where a minus sign means negative, zero means medium, and a plus sign means positive.

Power source comparisons			
	Batter-ies	Diesel	LP-gas
Capital cost	−	+ (lower)	+
Operating cost	+	+ (lower)	0
Life	+	0	−
Operating time between recharging/refuelling	−	+	0
Time expended on recharging/refuelling	−	+	0
Recharging/refuelling equipment cost	−	+	0
Recharging/refuelling — need for a specific location	+	−	0
Driving characteristics	0	0	0
Noise	+	−	0
Exhaust fumes	+	−	−
Vibration	+	−	−
Maintenance	+	0	0

Exhaust contents for different power systems

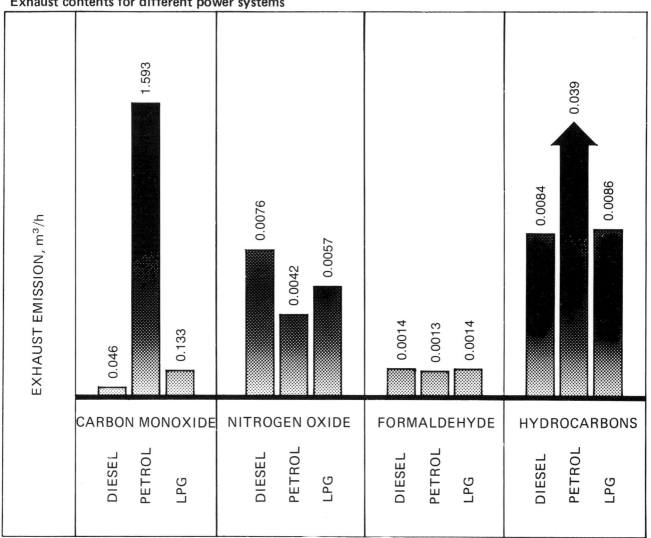

Bar chart showing tne carbon monoxide, nitrogen oxide, formaldehyde and hydrocarbon contents of exhaust fumes for diesel, petrol and propane powered engines.
Exhaust emission rates in m^3/hour for different fuels when operating a 2.5 tonne counterbalanced truck in short cycles.

Transmissions

With battery power, the motor is often linked directly to the traction wheels and no special transmission unit is required. With other power sources, transmission is normally by one of four different methods.

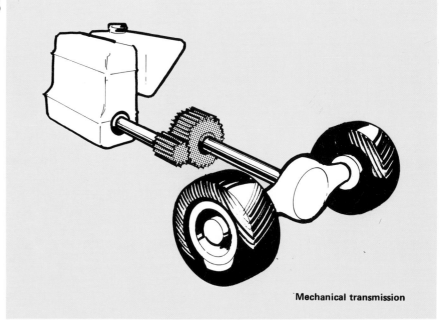

Mechanical transmission

Mechanical transmission results in high total efficiency. It generally requires three control elements (accelerator, clutch and gearchange) which often demand more work from the driver than do other systems. Mechanical transmission in trucks is now very unusual.

Hydrostatic transmission

With **hydrostatic transmission**, an electric motor drives a hydraulic pump which in turn drives one or several hydraulic motors connected to the wheels.

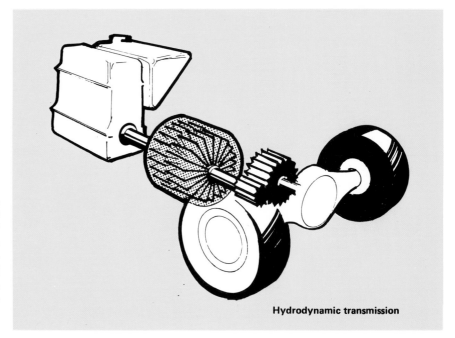

Hydrodynamic transmission

With **hydrodynamic transmission**, an electric motor drives a turbine wheel which in turn drives another turbine by means of a fluid (oil) coupling. Hydrodynamic transmission is frequently combined with automatic gearchanging to increase efficiency and is very common on IC trucks.

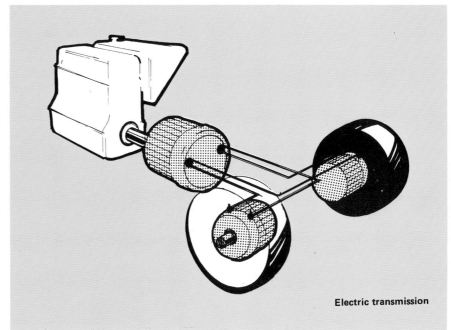

With **electric transmission,** an electric motor drives a generator which in turn drives one or several other motors linked to the wheels. Like hydrostatic transmission, it is very flexible. However, it is not used extensively.

Electric transmission

Wheels

As trucks often lack suspension, it is important that the wheels, and more often the tyres, are capable of deformation and resilience to protect drivers and loads. The less satisfactory the driving surface is, the greater the demand for these features. As resilience increases, so does rolling resistance.

Solid wheel

Solid tyre wheel

Solid wheel

Solid tyre wheel

Solid wheels are made from some hard material, frequently steel or nylon. They have low rolling resistance but require very even floor surfaces as well as low driving speeds. Solid wheels are long lasting and, obviously, puncture-proof. Steel wheels wear out floor surfaces.

Solid tyres consist of a solid natural or synthetic outer mounted on a rim. Rolling resistance exceeds that for solid wheels but floor surfaces need not be so even. Floor wear is lessened, but wheel life is shorter.

Pneumatic tyre

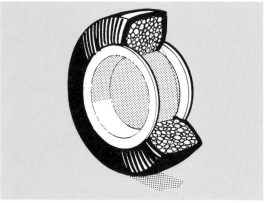

Semi-pneumatic tyre

Pneumatic tyres are inflatable (tubeless) and are mounted on rims. Great rolling resistance (and thus greater energy consumption) is experienced but the tyre makes less exacting demands on the floor surface. Pneumatic tyres may blow out and thus could represent a safety hazard.

Semi-pneumatic (cushion) tyres contain a number of air filled cells or cavities. Flexibility is less than that for pneumatic tyres but life expectation is greater. Cushion tyres are proof against blow-out, a valuable feature where the terrain is unsatisfactory, e.g. in scrap yards. Cushion tyres cost more than pneumatic tyres.

Pneumatic and cushion tyres are normally used out of doors.

Loading attachments

A loading attachment may be used simply to lift a load or to manipulate it, e.g., to invert it, to rotate a drum, etc.

The commonest attachment is a **fork** which consists of tines made of flat or sectional bar, bent at right angles and with tapered ends. Fork handling pre-supposes a load on a load carrier (usually a pallet) which elevates the load from the surface on which it rests.

Forklift truck.

Trucks are often used for extended periods to handle particular loads, e.g., paper bales or oil drums. To increase productivity or to enable trucks to accommodate loads which could not be managed with forks, a special **auxiliary attachment** may be fitted. Remember, though, that this may alter the lifting capacity and stability of the truck.

The ability to switch quickly from one attachment to another is important, and quick change fittings are often available.

Special attachments are used mainly on counter-balanced trucks.

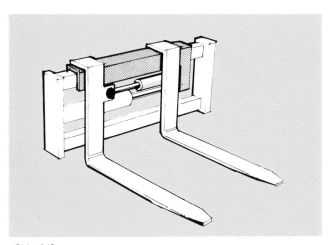

Side-shift.

Side-shifts are the most common auxiliary attachment since they enable the driver to save time by re-positioning the forks precisely in front of the load rather than moving the whole truck. Side-shifts should permit good driver visibility, otherwise the load may be damaged.

Truck handling lengthy goods, using fork-spreaders.

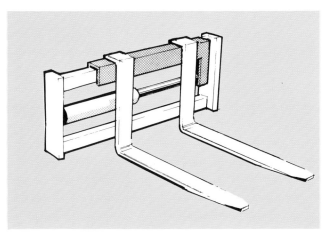

Fork-spreader

Truck equipped with side-shift.

Fork-spreaders change the distance between forks and often incorporate side-shifting. Fork-spreaders are needed when handling different types of pallet or extra wide loads.

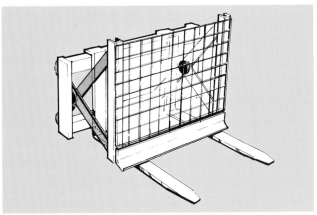

Push-pull attachment.

Truck fitted with push-pull attachment.

An alternative to forks is the **push-pull attachment.** Materials are positioned on a slipsheet (which, since it costs considerably less than a pallet, may accompany the materials and be kept by the consignee). Slipsheets take up little space and are also suitable for closed systems involving bulky materials. A drawback is that handling operations generally take longer.

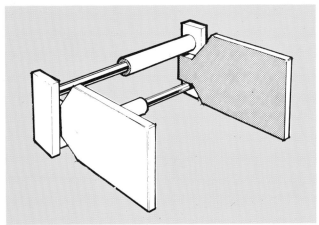

Clamp.

Truck fitted with clamp.

Clamps can be used instead of forks to handle cardboard boxes, paper bales, paper reels and other, similar materials. No base support is needed, which means that space that would otherwise be allotted to pallets or other load carriers is saved in warehouses, lorries, railway wagons, etc. The materials must be capable of tolerating the clamping pressure needed to lift them.

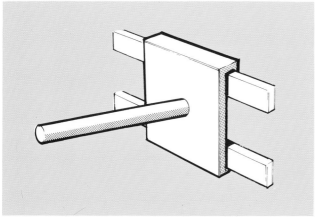
Boom.

Boom attachment handling.

Lengthy and/or cylindrical loads, such as coils of wire and rolled carpets, can be accommodated on a **boom attachment**. A boom offers increased handling capacity when compared with forks.

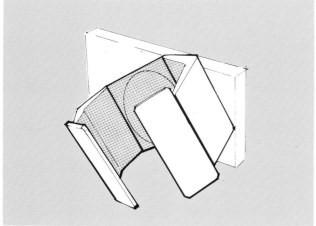
Rotating clamp.

Truck fitted with a rotating clamp.

Some loads need to be rotated, as when a drum is emptied or a paper reel is removed from a paper making machine. Instead of using a separate manipulator, the truck can be fitted with a **rotating clamp.**

Truck fitted with pantograph attachment.

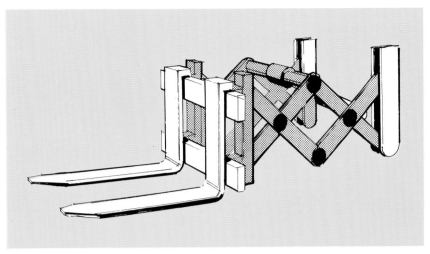

Pantograph.

Where the load position is some distance away from the nearest approach point for the truck (e.g., as with a lorry being unloaded from the side) a **pantograph attachment** is one answer.

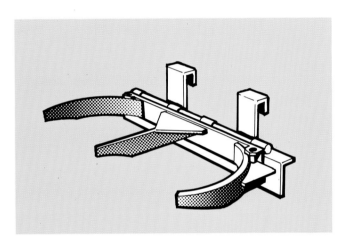

Drum-gripper.

Mounted drum-gripper.

Specialized units may improve handling rates and/or eliminate the need for a load carrier. In the diagram to the left, two drums are being lifted by a **drum-gripper;** no pallet is needed.

Utilization is improved if the truck serves as an all-purpose vehicle, i.e., can be fitted with snow buckets, refuse shovels, etc.

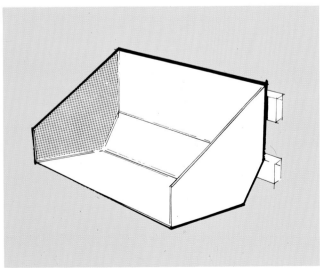

Bucket.

Truck equipped with dozer blade.

Lifting arrangements

A truck's lift mast consists of a number of frames sliding inside one another, usually on roller bearings. Lifting power is mostly provided by one or several hydraulic cylinders which move the sliding frames with or without the aid of chains.

Oil for the truck's hydraulics (to power lifting, auxiliary attachments and sometimes steering) is delivered by a pump. In battery powered trucks, this pump is driven by a separate electric motor, whereas in IC engine trucks there is a direct drive from the prime mover.

Four types of mast are commonly employed. A **single mast** consists of one frame. The fork carriage, with fork brackets, runs in the frame and provides mountings for the forks or auxiliary attachment.

For greater lift heights or with low-headroom trucks which are used inside containers, a second frame is added turning the mast into a **double (or duplex) mast**. Lift height may be nearly doubled while the closed mast height remains the same.

For even greater lift heights, there is a **triple (or triplex) mast**. **Quadruple masts** have been used but are rare. In these cases, one or two frames respectively are added.

Single, double and triple mast trucks.

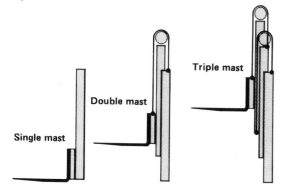

Mast operating principles.

In a single mast, the lift cylinder usually acts directly on the fork carriage. With multi-masts, the cylinder (or cylinders) acts on a sliding frame within the fixed frame. Uppermost on the movable frame is a guide pulley over which runs a chain. With duplex masts, the chain is fixed to the second frame. Third frames have a corresponding guide pulley and chain. In principle, any number of frames may be assembled in this manner, but in practice no more than three transmission stages (a quadruple mast) are used. The hydraulic cylinder power increases with the ratio of cylinder-to-carriage movement.

It is often desirable for a mast to have **free lift**, i.e., the forks are raised without increasing the mast and truck heights, as when working inside containers or railway wagons (see free lift height description under 'Performance').

Trucks with and without free lift, respectively.

Performance

Lift height

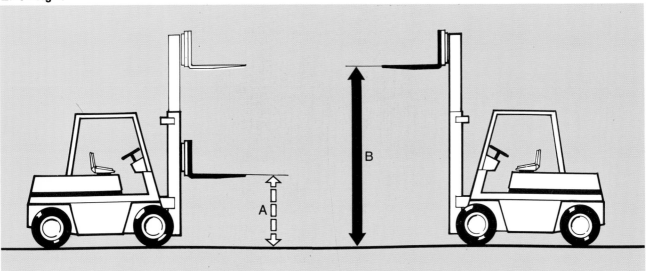

Lift height (A) and maximum lift height (B).

'Lift height' means the distance between the upper faces of the forks and ground level when the latter is horizontal and the truck mast is vertical. In everyday language 'lift height' actually means 'maximum lift height'.

Free lift height

In conventional mast operation, the frames extend, increasing the total truck height as soon as lifting begins.

Trucks with (right) and without (left) free lift.

With low headroom, the mast would hit the ceiling long before the load had reached the desired height. By adapting the mast height to the ceiling height, the load may be raised the maximum distance. On premises where the ceiling height varies (for instance, on board ships) or where the ceiling height is very low (for instance, inside containers), the truck must be capable of **free lift**.

Free lift height refers to the distance the load may be raised with no consequent increase in total truck height.

Fork tilt

Forward and backward fork tilt.

To prevent a load from sliding off the forks (for instance, when braking), most truck forks may be tilted backwards. On counterbalanced trucks tilting amounts to 10—12°. Trucks used indoors, especially reach trucks, travel more slowly and thus require only about 4° tilt.

Forks can also be tilted forwards to prevent them from getting stuck within the pallet. When picking up a load, the forks are tilted slightly forward, inserted under the load, re-tilted backwards and then lifted. Depositing a load involves the reverse procedure. Forks on outdoor trucks are normally capable of approximately 6° forward tilt and on indoor trucks 2—3°.

Tilting may be achieved by moving the entire mast or just the forks. Counterbalanced trucks usually employ mast tilting because of the greater tilt angles required. Reach trucks, which operate at great lift heights, often employ fork tilting to avoid stability problems.

Lifting capacity

In Sweden, lifting capacity is determined at a 600 mm load centre. This dimension refers to the distance between the centre of gravity of the load and the fork uprights. If the load centre is further out on the forks than the rated distance for the truck, lifting capacity is severely reduced.

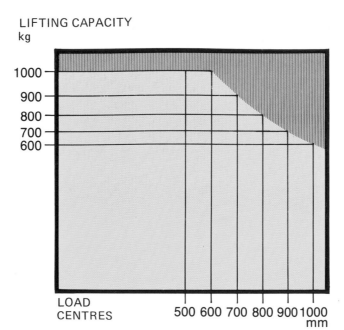

Correlation of load centre and lifting capacity.

Rated lifting capacity is the greatest load the truck can carry and still pass stability tests (SMS 2788, 2797 and 2986) at 2500 mm lift height (pallet base stackers) and at 3300 mm lift height (other trucks).

Maximum lifting capacity is the greatest load the truck can carry and still pass stability tests (SMS 2788, 2797 and 2896) at maximum lift height. Maximum lifting capacity is considerably less than rated lifting capacity.

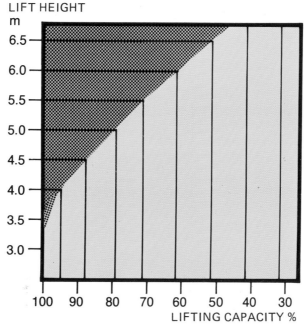

Correlation of lift height and lifting capacity.

Turning radius and aisle width requirements

A turning truck revolves round an assumed point. Turning radius is the radius of the smallest circle having this point as its centre and encompassing all parts of the truck.

The aisle width requirement is calculated according to Swedish Standard (SMS 2799) as the width of the aisle through which a truck is able to reverse in a straight line with an 800 mm wide and 1200 mm long load and then turn through a right-angle (i.e., to deposit or pick up a load), maintaining a margin of 100 mm between the truck (and/or its load) and fixed points such as racking, building supports, other loads, etc.

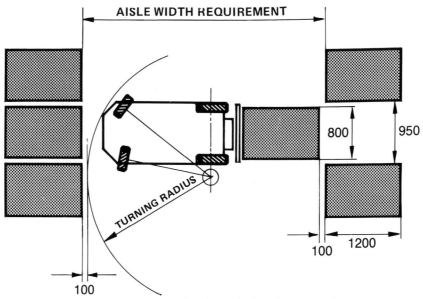

Turning radius and required aisle width for a four wheel truck.

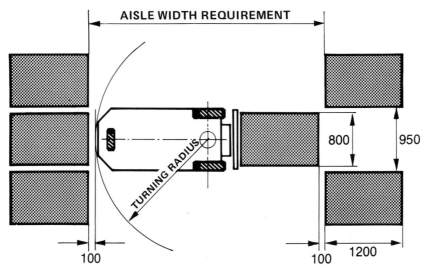

Turning radius and required aisle width for a three wheel truck.

Stability

Trucks sold in Sweden must conform to stability requirements specified in Swedish Standards (SMS 2788, 2797 and 2986).

Different tests are used for different trucks i.e., high-lift counterbalanced trucks, high-lift trucks with retractable load carrying attachments, and high-lift stackers.

Gradeability

The longer an electric motor works under a given load, the more its power output diminishes. The gradeability of a battery powered truck is indicated in two ways.

Simple **gradeability** is the percentage incline (dry and even surface) on which the truck is able to stop and re-start, with the motor producing an output equivalent to that which it can sustain for a thirty-minute period.

Maximum gradeability is when the motor produces an output equivalent to that which it can sustain for five minutes. In both cases the truck must be able to reach a speed of 0.6 m/sec.

For an IC engine truck, gradeability is indicated as the greatest percentage incline (even, dry surface) the truck can surmount over an extended period. Even here, the truck must be able to stop, re-start and reach a speed of 0.6 m/sec.

Gradeability and maximum gradeability are indicated with or without load.

Truck drivers and their working environment

For drivers, the most important features of a truck are visibility, seat design and noise level.

Visibility

There is as yet no officially recognized method of measuring visibility (1980) but such a method is likely to be devised within a few years. Detailed visibility requirements may then be included in the Swedish International Safety Board's Directive No. 94, 'Ride-on truck regulations'.

Visibility is divided into four types: driving visibility, close visibility, fork tip visibility and fork heel visibility.

Driving visibility is important when the truck is driven a considerable distance at high speed. The driving direction considered to be normal varies for different truck types.

A counterbalanced truck may be driven backwards as often as forwards if large, bulky loads are being moved. Whatever the driving direction, the driver should be able to see clearly.

Driving visibility includes side visibility — of importance when the truck reaches a crossing. It may be acceptable to have the driver move quite considerably to achieve the necessary degree of side visibility. But acceptable front or rear visibility should be possible with the driver seated normally. Where the driver faces forward and has to reverse some considerable distance he may find it difficult to turn his body sufficiently. For such cases, rear visibility may be improved by fitting a pivot seat or by positioning the driver so that he faces sideways.

Close visibility is essential when the truck is picking up or depositing loads in confined spaces, etc.

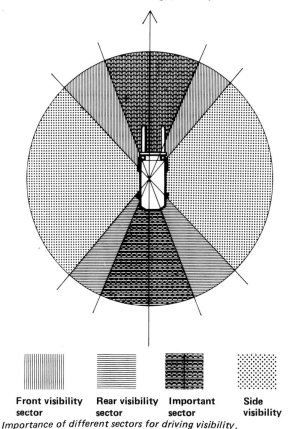

Front visibility sector   Rear visibility sector   Important sector   Side visibility

Importance of different sectors for driving visibility.

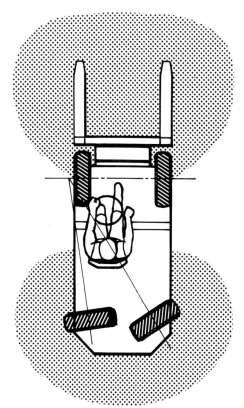

Importance of different sectors for close visibility.

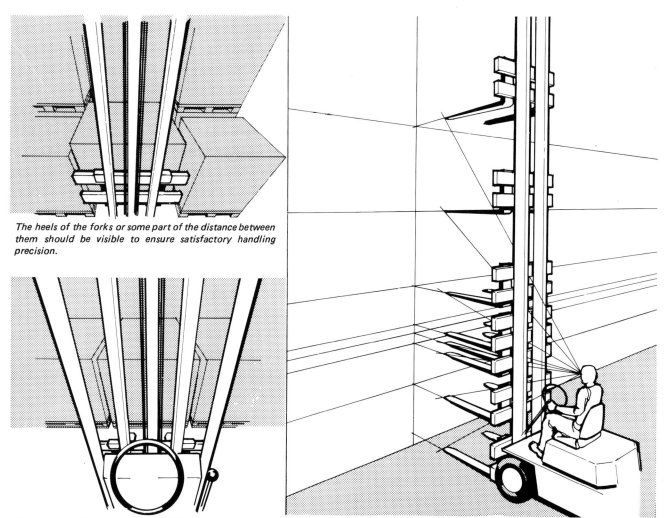

The heels of the forks or some part of the distance between them should be visible to ensure satisfactory handling precision.

At least one fork tip must be visible when picking up a load in order to achieve efficient handling without damaging the materials

The driver should be able to see at least one fork tip at all applicable handling heights without having to move his body excessively.

As such work frequently involves turning, good visibility will be required on the front and rear 'blind' quarters and in other sections close to the truck. In assessing close visibility it should be remembered that the driver moves quite considerably.

Good-fork tip visibility enables the driver to remove or deposit loads efficiently and safely. Where the truck is fitted not with forks but some other attachment, the visibility requirements for that attachment must obviously be assessed. When a truck is used solely for handling pallets on an even floor surface, often only one fork tip need be visible. When pallet loads vary in shape, both tips should be visible. Fork tip visibility is required when serving lorries (1.2 to 1.45 m), conveyors, and at all other operating heights.

Fork heel visibility is important when depositing a load, say, in pallet racking, on a lorry, on the floor and especially — when handling vulnerable materials. It is also necessary when a truck is used to rest one pallet on another. Either the actual fork heels or part of the distance between them should be visible and should be assessed in relation to all operating heights (as with fork tips).

To summarize, general driving visibility and close visibility are important for safe and efficient fork truck operations. Fork tip visibility is an important aid to efficient handling and to prevent damage to materials. Fork heel visibility is vital for satisfactory handling precision.

The visibility problems connected with trucks are described in greater detail in TFK Report 1978:7, 'Visibility requirements for fork lift trucks' which also recommends a method of measuring visibility.

Body position

To ensure that a fork truck driver's body position is correct, the seat must be well designed, the controls well placed, and the field of view must be satisfactory without the need for excessive body movements. The seat should be stable and spacious, provide a comfortable sitting position, and should be adjustable as required. Where the truck has to be driven on uneven surfaces which cause vibration, the seat should have some form of suspension. On counterbalanced trucks, the driver sits facing forward but frequently has to reverse. He must therefore be able to position himself on the seat at an angle which precludes excessive neck twisting. For that reason the seat must not be too cupped. 'Ride-on truck regulations' indicate general seat requirements.

Noise

'Ride-on truck regulations' indicate how noise is to be measured, and commercially available trucks should meet the requirements stated. Basically, drivers should not be exposed to a sound level exceeding 85 dBA; this corresponds roughly to traffic noise in a street.

Where trucks are equipped with cabs, noise regulations apply both when windows and doors are open and closed.

Comfort

'Ride-on truck regulations' prescribe that truck cabs should be well heated, insulated and capable of ventilation. Trucks without cabs used in cold conditions should have heated seats.

Draught in cabs should be considered. Some air conditioning units are so draughty that drivers are disinclined to use them. Instead, they open a window causing dust and noise to enter the cab.

Training

'Ride-on truck regulations' require truck driving personnel to be well acquainted with the operation and use of their vehicles. This means that, after completing a truck driving course, drivers should have passed a test covering theoretical and practical aspects. Training information is available from the Swedish Transport Trades' Occupational and Working Environment Committee (TYA) in Stockholm or from major truck suppliers who also offer courses based on National Swedish Board of Education syllabuses.

Training should cover a wide range of matters. For example, maintenance may be a driver's responsibility. Other possible subjects are regulations governing truck use, internal traffic regulations, driving regulations, loading and unloading provisions, damage, accidents, etc. Swedish Standard SMS 2793 indicates how drivers' instructions should be drawn up.

Truck driver's badge.

Truck drivers should also be at least eighteen years old and should have the management's official permission to drive trucks. They should preferably also wear a badge showing they are entitled to drive trucks.

Counterbalanced trucks

3 tonne counterbalanced truck.

Counterbalanced trucks carry their loads outside the stability polygon — an area limited by assumed lines drawn between the truck's points of contact with the floor/ground.

The load's tendency to tip the truck forward is counteracted by the truck's counterweight. To increase the load capacity additional ballast is normally mounted at the rear of the truck. Hence 'counterbalanced truck'.

Counterbalanced trucks are the most common truck type and can be used for handling most forms of materials. Some 40 to 50 per cent of trucks sold in Sweden are counterbalanced trucks and have either *three* or *four wheels*.

Three-wheel battery powered counterbalanced truck.

Four-wheel battery powered counterbalanced truck.

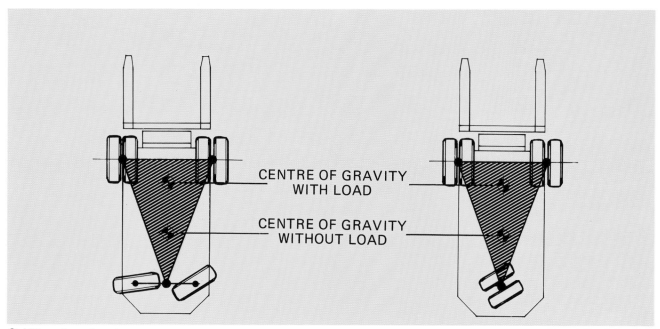

Stability polygon for a four-wheel truck.

Stability polygon for a three-wheel truck.

The stability polygons are equal for both forms of truck. This is because four-wheel trucks are equipped with a pivoted steering axle, i.e., the axle pivots at a central point on the truck.

Four-wheel trucks negotiate uneven driving surfaces with greater ease and are usually chosen for such application.

Aisle width requirements for a three-wheel truck are less as is its turning radius. Three-wheel trucks are often battery powered. They seldom exceed 1—2 tonnes (of rated lifting capacity). Four-wheel trucks, however, are available in all sizes from 1—30 tonnes.

Application

Four-wheel counterbalanced truck loading a lorry.

Large four-wheel counterbalanced truck handling a container.

Three-wheel counterbalanced truck.

Medium sized four-wheel counterbalanced truck.

A counterbalanced truck handling paper reels.

Three-wheel counterbalanced truck stowing goods in a container.

As counterbalanced trucks always carry their loads outside the chassis area they can accommodate a large range of items although the size should not be excessive. They are useful as general purpose machines and are employed in most branches of trade and industry.

In some instances more specialized trucks or equipment may be superior. A counterbalanced truck is large and heavy compared with other truck types of equivalent lifting capacity. It takes up more space and may possibly be uncompetitive with a reach truck in operations involving indoor stacking in pallet racks. Its weight and size may also make it slower and less manoeuverable than other, smaller truck types.

Counterbalanced trucks can accommodate good sized cabs and comparatively large wheels and are therefore suitable for outdoor use on uneven ground.

Counterbalanced trucks are chosen when:

- There is general purpose work and the need to manage many different types of materials, possibly with the aid of auxiliary attachments.
- The work is out of doors under cold conditions or over rough surfaces.
- Lorries and railway wagons have to be loaded/unloaded from ground level.

Counterbalanced trucks are **less suitable**:

- For materials handling in confined spaces, e.g., indoor warehouse stacking with stillages.

Counterbalanced trucks are often used for:

- General purpose work in light industry, warehouses, depots, etc.
- Harbour work (in various sizes) involving general cargo and containers associated with warehouses, ships, lorries and railway wagons.
- Timber yards (timber bundles, etc.).
- Paper reel and pulp handling.
- Transportation between processing and warehouses in the engineering industry.
- Work at lorry and railway terminals.

Design

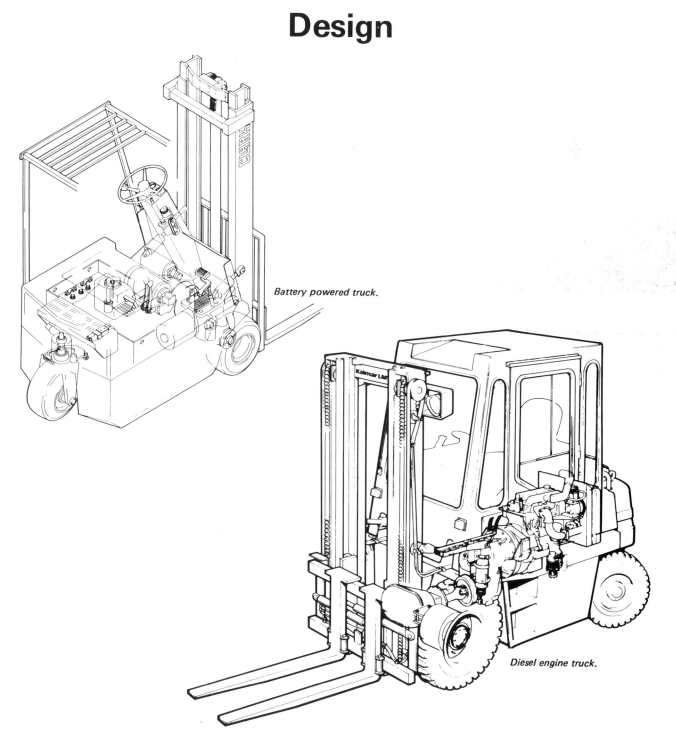

Battery powered truck.

Diesel engine truck.

Primary truck functions are described earlier. The section below deals only with aspects peculiar to counterbalanced trucks.

The **mast** may incorporate one, two, three or four elements, depending on the required lift height and permissible total height. The mast supports the fork carriage on which the load carrying attachment is mounted. The most common attachment is a set of forks. The mast pivots at the truck chassis so that it may be tilted by cylinders mounted in the mast and chassis. Mast or fork tilting is required to ensure that:

- the load remains on the forks during transport (backward tilt),
- the truck can pick up and deposit loads more easily (forward tilt).

Normal mast tilt angles for counterbalanced trucks are approximately 6° forward and 10–12° backwards. The auxiliary attachments described earlier can all be used on counterbalanced trucks.

Mast design determines such properties as lift height, and manoeuverability in confined spaces.

Since the driver sits behind the mast he must also be able to see clearly through it. Visibility has an important influence on safe driving and on the truck's ability to pick up and deposit loads. Truck visibility is discussed generally above.

To improve visibility, clear view masts were introduced during the late 1970s. Lift cylinders, chains and hoses were removed from the centre of the mast and incorporated in the mast uprights. Although visibility through the mast has improved, it has deteriorated slightly in the sectors obscured by the mast uprights, lift cylinders, chains and hoses. Mast design is also more complex and expensive.

A lift cylinder approximately half as long as normal and incorporating a two-stage telescopic function, also improves visibility to the front. Driving visibility is much better in comparison with that for conventional masts, but fork tip and heel visibility are slightly impaired. In terms of visibility, therefore, the former design is preferable to the latter. The different mast types are shown below in three typical designs.

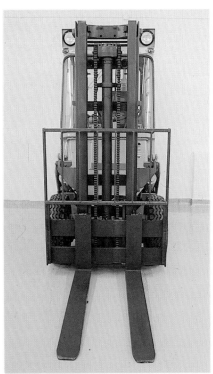

Truck with a conventional mast.

Truck with a clear view mast and one cylinder.

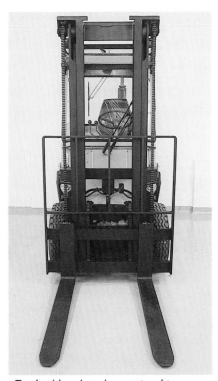

Truck with a clear view mast and two cylinders.

Bigger counterbalanced trucks (over 7 tonne capacity) are nearly always diesel powered. Smaller trucks may be powered by diesel engine, LP gas or battery.

Four-wheel trucks are often fitted with a pivoted steering axle mounted at the centre of the truck. Drive is usually to a rigid axle. On three-wheel trucks the steering axle is replaced by a steerable wheel. The third wheel — sometimes a single wheel, sometimes a pair side by side — may be used for both driving and steering. Although the tyres are frequently pneumatic or cushion, solid tyres are sometimes used (see previous text).

Steering is usually hydrostatic (servo assisted) which reduces the effort required from the driver. Hydrostatic steering permits the use of a narrow steering column which enhances visibility.

Most modern diesel-powered counterbalanced trucks are equipped with hydrodynamic transmission (automatic gearbox). Gearchanging is by means of a small lever, frequently servo assisted and easy to operate. There are normally two or three gears forward and reverse.

The rigid front axle is of a heavy duty design as it carries almost the entire truck deadweight plus the weight of the load. Counterbalanced trucks experience high wheel pressures in relation to the loads carried (front axle pressure = double lifting capacity).

A fully-laden truck has lighter steering which can cause problems on slippery surfaces. On the other hand, when the truck is driven without load, stability problems can occur. The fact that these trucks nearly always lack suspension further impairs driving characteristics. There is every incentive to drive trucks carefully irrespective of whether or not they are carrying loads.

Trucks used out of doors or within unheated premises during the colder months are often equipped with cabs. Otherwise, they have only a load guard above the driver (demanded as protection against falling materials by 'Ride-on truck regulations').

To achieve a body position satisfactory for reversing, the driver must be able to turn round in situ. **Swivelling seats** have been installed on large trucks and **angled seats** on small trucks. Rotation of 45—60° is often sufficient for a swivel seat. Where the seat is fixed, the cupping must not prevent the driver from taking up an angled position. The **cab** (where applicable) often has large glass windows and these should be fitted with windscreen wipers, front and rear. Frequently, there is a roof window enabling the driver to see an elevated load. This window, too, should have its own wiper.

Another important feature is the design and ease of operation of the **controls**. Unhindered slow and rapid lifting, lowering and mast tilting should be possible.

The cab step is particularly important on small counterbalanced trucks where the driver frequently has to leave his cab in order, say, to re-arrange materials incorrectly positioned on the pallet. There should be no possibility of slipping, for instance, on an icy step. Door handles should work from inside the cab as well as outside. Door windows should be easily opened and closed so that the driver can, for instance, talk with a supervisor or reach a switch to open a gate.

Silencers demand particular attention. The Swedish Industrial Safety Board requires that the sound level in cabs should not exceed 85 dBA, but today (1980) many trucks have cab sound levels far below this figure. This is fully justified since sound insulation gradually deteriorates.

3 tonne diesel powered counterbalanced truck.

Maintenance

Regular maintenance prolongs truck life. On electric trucks, for example, the battery level must be checked and, if necessary, topped up daily. One method of reducing this chore is the use of sealed **batteries** which are not subject to evaporation. Alternatively, trucks may be equipped with a central system by which levels are checked and water is added from a single point.

If a diesel powered truck is used extensively, oil level should be checked daily. Hydraulic fluid should also be checked fairly often. Articulated bearings, e.g., steer axle suspension, mast mountings and all hydraulic cylinder mountings, should be lubricated occasionally. Filters for engine oil and hydraulic fluid will need changing. All such tasks must be easy to perform. Big machines may have central lubricating systems to facilitate maintenance. Similar systems may eventually be installed even on small trucks.

Future developments

Although there are no reliable statistics, the number of counterbalanced trucks in use in Sweden is estimated at 20,000. Battery operated trucks have tended to grow in capacity and today are produced with lifting capacities of up to 6 tonnes.

There is a demand for large trucks to be used indoors on unventilated premises. Even small battery powered trucks need greater capacity and energy resources to manage auxiliary attachments (such as clamps) and cab heating. In the long term, a new type of battery may solve the problem. Hydrogen power represents another possibility. This development is described in greater detail in TFK Report 1979:2, 'Alternative truck power sources'.

Masts may be further developed. As mentioned earlier, conventional masts have been superseded more and more by clear view masts. But even these may provide insufficient visibility for some — primarily big — trucks where uprights and cylinders combine to limit visibility wherever they are positioned. Future improved lifting arrangements may involve completely new designs.

In certain industries, counterbalanced trucks are getting bigger and bigger, mainly because of increases in the size of the loads. In the pulp industry, for example, units comprising 4 to 8 bales used to be handled. Nowadays 16 and 32 bale units are common.

Truck handling a 32 pulp bale unit.

In some areas, for instance, users who previously chose 4–5 tonne trucks now often choose 7–8 tonne trucks. This trend is likely to continue.

The use of **driverless trucks** will increase. Quite conventional counterbalanced trucks are nowadays controlled by electric loops. Alternatively, the trucks may have a driver. The fork tips have sensors capable of identifying the slots in a pallet. Such a truck can pick a pallet automatically from a given level, transport it and deposit it at a different level. In this way, the truck becomes an alternative to a conveyor. Other, more specialized types of driverless truck are called automatically guided vehicles (AGVs). See special chapter.

In some countries, the price of counterbalanced trucks has increased greatly over the last few years. This is even true in absolute terms. The increases are due above all to the need to meet improved environmental requirements. Sound insulation, to take one example, has become a comparatively expensive part of the cost of a diesel truck.

Traction batteries for battery powered trucks now cost much more. Although price increases will not be as great in the future, they are likely to continue. Assuming a truck life of 8–10 years, capital costs today account for a very small part of the total (which also includes drivers' wages, maintenance, fuel, tyres, ventilation and sound insulation).

Powered stackers

Powered stackers carry their loads mainly inside the stability polygon to give good space utilization. They are more compact than counterbalanced trucks and can operate in narrower aisles.

Powered stackers come with different fork concepts: pallet base and straddle base.

Pallet base stacker.

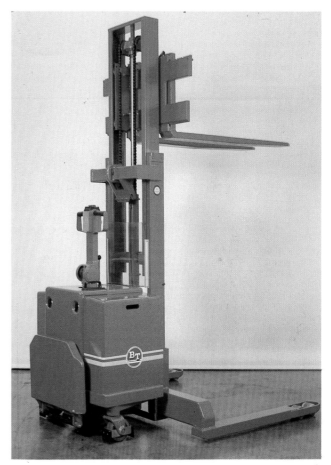

Straddle base stacker.

Fixed wheels — often solid or with solid tyres — are mounted in the support arms of **pallet base stackers**, and the forks envelop the arms from above. Both arms and forks are inserted beneath the load.

The support arms of the **straddle base stacker** are adjacent and parallel outside the forks. When fully lowered, the forks (and therefore the lower surface of the load) are below the upper surface of the support arms. The support arm wheels may be larger than those for the pallet base stacker and the forks thinner. Larger wheels are better able to cope with uneven surfaces. In any case, straddle stackers are more stable than pallet base stackers.

Powered stackers may be:

- High lift (e.g., for stacking to several levels of racking as well as for transport).
- Pedestrian guided (the operator walks beside the truck to steer it).
- Ride-on (the operator drives the truck).

Powered stackers are used primarily where space is limited (e.g., transport in warehouses, workshops, rail wagons, containers, etc.). Because they are comparatively cheap, they are very common. As a result, they are used for tasks other than those they were designed for.

Stand-on powered stacker with fold-up operator's platform.

Sit-on stacker.

Pedestrian and ride-on powered pallet trucks

Application

Pedestrian operated pallet trucks are the simplest and cheapest powered vehicles for handling pallets and stillages. They are designed to transport materials over short distances (i.e., less than 50 metres) on hard, smooth floors. Because they are controlled by pedestrian operators, their maximum speed is set at 6 km/h (1.67 m/sec). Where a folding ride-on platform is fitted, longer distances are possible but still at a maximum speed of 6 km/h. These trucks are very compact and are thus suitable for use where space is limited. Lifting capacity is 1,000—2,000 kg and maximum lift height approximately 0.2 m, i.e., sufficient only for ground-level operations.

Pedestrian and ride-on pallet trucks can handle Euro pallets and roll pallets. The forks are comparatively thick, and short-side handling, therefore,

Order picking is a common application. The driver can mount and dismount with ease. Where picking distances are reasonably short, the slow driving speed is no disadvantage. Specialized picking trucks are available and are described in a separate section.

Loading and unloading of lorries and railway wagons are other important applications where these powered trucks are an alternative to simple pallet trucks.

Ride-on pallet trucks are used for long distance ground level operations, for instance, between goods reception and goods marshalling areas. They are fast and their carrying capacity is great.

is the most common application. For long-side handling the fork wheels must pass fully through the pallet (so that it can be lifted without risk of breaking).

Stand-on and — more especially — sit-on powered pallet trucks are better than pedestrian versions for moving materials long distances. Driving speed is higher, normally 6—8 km/h (standing drivers) and 8—12 km/h (seated drivers). Maximum speed with the driver standing is 16 km/h, otherwise there are no restrictions. Ride-on pallet trucks are slightly bigger and more expensive than pedestrian pallet trucks and are less commonly used for loading and unloading lorries and railway wagons.

Design

Powered pallet trucks come in four types depending on the positioning of the operator.

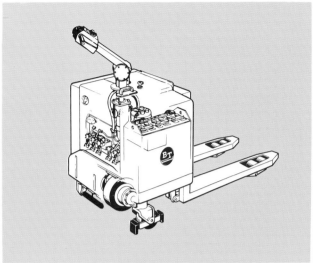

Pedestrian operated powered pallet truck.

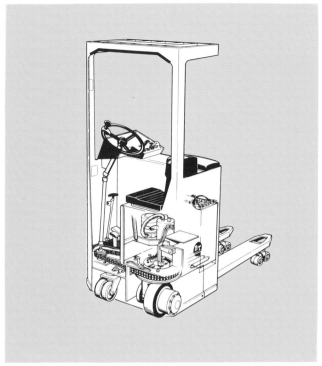

Sit-on powered pallet truck.

Powered pallet truck.

Stand-on powered pallet truck.

Powered pallet truck with fold-up stand-on platform.

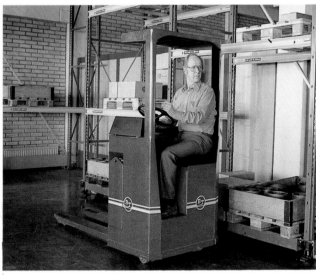

Sit-on powered pallet truck.

In all four cases, the support wheels are mounted via a linkage under the forks and are often referred to as 'fork wheels'.

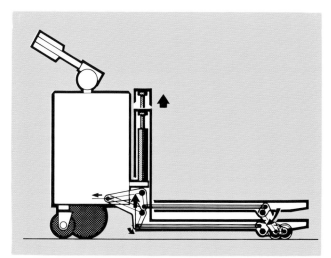

Lifting principle of pedestrian powered pallet truck.

A hydraulic cylinder lifts the fork unit relative to the chassis. As the linkage is actuated, the fork wheels are pushed downwards, raising the forks at both ends. The movement is such that the forks remain horizontal during the operation.

Because of the size of Euro pallets, fork wheel diameters are limited to approximately 85 mm. Floors must be firm and even; door sills should be non-existent or very low. Fork wheels are often mounted in a bogie for smoother running on uneven surfaces and reduced wheel pressure. Fork wheels are often made of steel or nylon, or of either of these materials covered with natural or synthetic rubber. The two latter materials inflict the least wear on floors, steel the most. Natural

rubber has the lowest strength.

With all these trucks, the driving wheel also serves as a steer wheel. The most common of the two mounting methods (see illustrations) involves a centrally-located, spring-loaded drive/steer wheel, with support wheels on either side. The support wheels are free-swivelling castors and do not affect the steering.

With the second method, the drive/steer wheel is mounted at one side, with a support wheel opposite. In this case, the drive wheel is not spring-loaded. The support wheel is again a free-swivelling castor. Both the drive/steer wheel and castor carry the truck. The drive/steer wheel is operated by means of a tiller (see figure) or steering wheel. The castors require some kind of guard to protect feet from being run over.

Ride-on trucks are controlled either by side-facing seated drivers or standing drivers. The controls are typically automotive: steering wheel, accelerator and brake pedals, and a device for selecting travel direction. Standing drivers stand on one side of the truck. Steering is by a steering wheel or a short lever. The 'dead man's brake' is applied when the pedal — depressed during operation — is released. The other controls are hand-operated.

The trucks are battery powered as they are normally used indoors. Battery powered trucks are described in greater detail in an earlier section.

The drive motor is generally connected direct to the combined drive/steer wheel. A hydraulic pump driven by a separate electric motor provides lift.

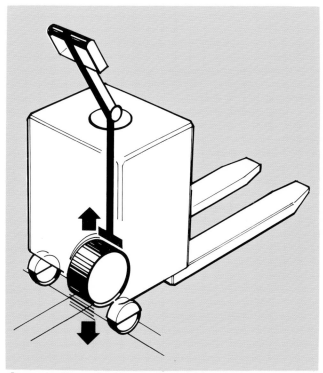

Powered pallet truck with spring-loaded drive wheel.

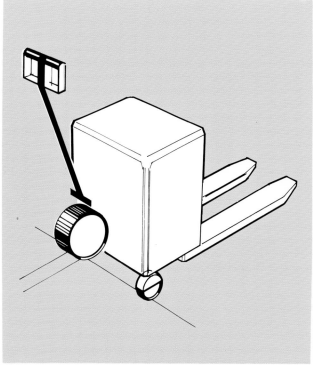

Side-mounted drive wheel is not spring-loaded.

Pedestrian and ride-on pallet base stackers

Powered stackers are divided into four different basic types depending on the operator's position.

74

Pedestrian powered stacker.

Stand-on powered stacker with fold-down platform.

Ride-on powered stacker with a standing driver.

Ride-on powered stacker with a seated driver.

Application

Stacker trucks are used where there is a need for low to medium volume moving and stacking of pallet loads and where the floor surface is satisfactory.

The truck's support arms require free entry at floor level and pallets must be positioned accordingly. Loading and unloading lorries from ground level is impeded at the points where the lorry wheels prevent close access.

Maximum permitted speed for a pedestrian stacker is 6 km/h (1.67 m/sec). For ride-on stackers the permitted speed is 6–8 km/h (standing driver) and 8–12 km/h (seated driver). Maximum permitted speed for vehicles with a standing driver is 16 km/h, otherwise there are no limits. Lifting capacity is normally 1,000–1,500 kg. For reasons of stability, the maximum lifting capacity normally can be utilized only up to 2.5 m for pallet base stackers and 3.3 m for straddle base trucks. Maximum lift height is approximately 4–5 m where trucks are not fitted with stabilisers, and approximately 6 m when these are fitted (see page 76).

Low and medium volume stacking of pallet loads is a common application.

Ride-on stackers are often used for combined stacking and moving, e.g., replenishing assembly points with materials.

Design

Pedestrian powered stackers have support arms which are rigidly fixed to the truck chassis. The mast may be a single, double or triple mast and it executes the entire lifting movement. With pallet base stackers, the foldover forks consist of inverted channels which can be lowered over the support arms. More detailed mast information is to be found in an earlier section. The drive/steer wheel and the support castor(s) may be positioned in two different ways exactly as with powered pallet trucks.

Clearance beneath the support arms is comparatively low and may constitute a limitation when driving over loading ramps or on inclines. To increase the clearance, the stacker may be fitted with the same type of lifting support arms as those of powered pallet trucks.

Pallet base stackers are commoner than straddle base stackers.

Pedestrian powered stacker with lifting support arms.

Pedestrian powered stacker with load guard.

Ride-on Truck Regulations (ASS Directive 94) prescribe an overhead guard where there are risks of goods falling. There are no corresponding regulations for pedestrian trucks (1980). An alternative to an overhead guard for pedestrian trucks is a load guard mounted on the fork carriage to prevent goods falling backwards.

Straddle base powered stacker.

Ride-on powered stacker with stabilisers.

To increase the stability of pallet base stackers, stabilisers may be fitted. These are used when stacking at greater heights, but as it takes time to engage them, their application is in low volume work.

In most cases, both lift and drive are electrical. Hand-operated lifting mechanisms also exist. More information on battery powered drives is found in a previous section.

The support arms on straddle stackers are widely spaced so that they encircle the pallet on two sides, thus promoting high-lift stability. Thin, forged forks are used, and the distance between them may be varied to accommodate non-standard pallets. Larger wheels may also be fitted, permitting operations on slightly less satisfactory surfaces.

Future developments in powered stackers

In the Swedish powered truck market, only counterbalanced forklifts outsell stackers. Approximately 1,800 powered stackers and 2,000 counterbalanced trucks are sold annually.

The use of powered stackers is likely to increase because they will replace manual labour in a variety of operations. This especially applies to heavy lifting jobs, for instance, in shops, where a simple pallet base truck is beneficial. Another area of development is loading and unloading lorries and railway wagons. In manufacturing industry and the retail trade there is further considerable potential.

Reach trucks

Reach trucks combine the characteristics of counterbalanced trucks and powered stackers.

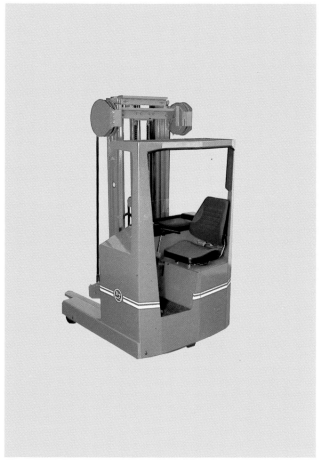

Conventional (moving mast) reach truck.

Pantograph reach truck.

Conventional (moving mast) reach trucks

With the mast retracted, a reach truck is, in essence, a straddle base powered stacker. When the load (often a pallet load) is lowered between the support arms, the truck is stable and requires only a small area for manoeuvre. With the mast extended, it functions as a counterbalanced truck. Loads of various type can be handled outside the chassis area and it is not necessary for the support arms to enter the racking as it is with a pallet base stacker. Because of their versatility these trucks are common in palletized warehouses. They stack in confined spaces and perform moving and loading/ unloading operations from floor level. As a result they are able to replace several other truck types.

Application

Reach trucks are designed primarily for pallet racking operations at greater heights than those suited to counterbalanced trucks. They require little space and are fast. Because they incorporate the advantages of counterbalanced trucks, they are also versatile. For instance, they can load and unload lorries and rail wagons from ground level.

Floors should be even, dry and clean; their load bearing capacity should be acceptable.

Reach trucks should also be used on heated premises as they normally lack cabs (although some versions — intended for use in cold stores — have cabs with LP gas heating). As stacking heights increase, floor evenness becomes more important.

Lifting capacity is normally 2,000 kg and maximum lift height 6—7 metres.

The main alternatives to reach trucks are ride-on powered stackers, counterbalanced trucks and pallet base trucks (see respective sections).

Ride-on stackers have less capacity than reach trucks and are not as versatile. They are best suited to low or medium height operations.

Counterbalanced trucks are used primarily out of doors or on unheated premises; loads are usually heavier than those typically handled by a reach truck. Counterbalanced trucks impose less wheel pressure than reach trucks — an important factor when the truck has to enter a container.

Pallet base trucks compete with reach trucks where the storage environment is very confined. Materials should be moved to and from such areas by some other type of truck, e.g., a ride-on powered pallet truck, reach truck or counterbalanced truck.

Design

Reach trucks are used as all-purpose vehicles in warehouses and stores. The 2,000 kg capacity reach truck shown can lift up to 6 metres and its forks can be tilted.

Reach trucks are also useful for loading and unloading lorries from ground level.

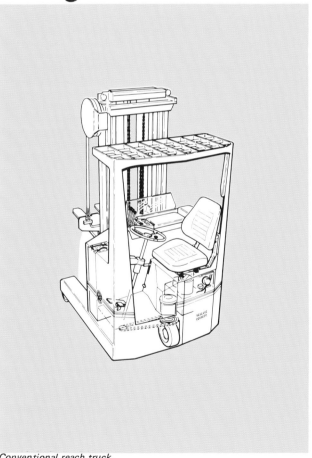

Conventional reach truck.

The mast on a conventional reach truck is similar to that on a counterbalanced truck. More details on masts can be found in the chapter 'General information on trucks'.

Reach trucks nearly always have side-facing driver's seats. On the whole, the control layout is like that of a car, with an accelerator pedal, brake pedal and steering wheel. There are three controls for moving the load: lift/lower, forward/backwards fork tilt and mast extension/retraction. Controls should be designed to avoid jerky movements. Facing sideways, the driver enjoys good forward and backward vision. The normal driving position, therefore, involves less strain than that for a counterbalanced truck. Because of this and the fact that reach trucks usually travel over fairly even surfaces, it is unnecessary to fit special seat suspension or shock absorbers.

Load guards are necessary as is a foot switch which must be kept depressed by the driver's left foot during driving. This ensures that the driver cannot dangle his foot outside the truck where it might be crushed.

Because the driver sits sideways, he can see past the mast when driving forward and does not need a clear view mast. Pallet widths do not differ widely and therefore it is sufficient if the driver can see one fork or one side of the pallet.

The trucks are battery powered — a subject covered in greater detail in the section 'General information on trucks'. The batteries power the traction motor and the hydraulic pump for most operations. The traction motor is connected to the combined drive/steer wheel under the truck (double drive/steer wheels are also employed). Steering movement in excess of 180° is possible, steering force being transmitted mechanically by a chain or hydrostatically (servo-assisted steering).

Wheels (usually cushion tyred) are fitted to the support arms. The latter should incorporate some form of guard to protect pedestrians' feet.

The mast is mounted on a carriage and moved in guides in the support arms by a hydraulic cylinder. Bearings must be free from play and sturdy as even minor movements will be considerably enhanced at greater mast heights. Bearing pressure is great and wear may occur with time. Some trucks have replaceable wearing surfaces to rectify wear problems.

If the mast is rigidly mounted to the carriage, fork tilting is achieved by pivoting the carriage with a hydraulic cylinder. Alternatively the mast itself may be pivoted. Mast tilting permits greater horizontal load movement at extended heights. As a result, loads can be deposited at great heights without moving the truck.

To ensure that mast deflection at great heights is not transmitted to the truck, causing it to topple over, the mast tilting mechanism should be jerk-free.

Trucks with tilting forks are more stable and full tilt can be utilized during the entire lifting motion. Another advantage is the increased precision of positioning, reducing the risk of pallets being deposited at an angle in the racking and subsequently falling to the floor. Where the mast is fixed, the whole truck has to be moved to position its load.

Provided the truck driver is experienced, neither tilting method has any speed advantage over the other.

Normal fork tilt angles are approximately 2° forward and 4° backwards.

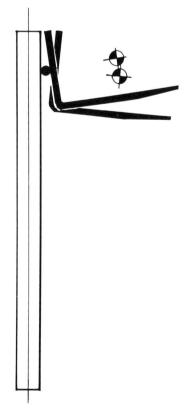

Centre of gravity shift with fork tilting.

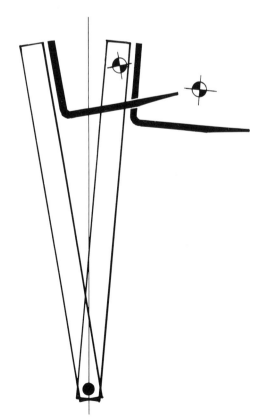

Centre of gravity shift with mast tilting.

Pantograph trucks

Pantograph reach trucks have the same range of applications as conventional (moving mast) types. The difference between the two is in the way the forks are extended.

A pantograph mechanism demands a more robust mast to yield the same performance since the centre of gravity of the extended load shifts a long way from the mast. In practice, therefore, pantograph trucks are not built to operate at maximum lift heights.

The advantage of a pantograph truck is that it can block stack where there is room for the support arms. This characteristic may also be utilized when loading and unloading lorries from ground level (from one side).

Pantograph trucks are rare in Europe but common in the United States — mainly because US warehouses are often built with lower racking.

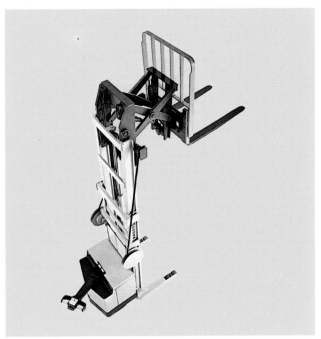

Pantograph truck.

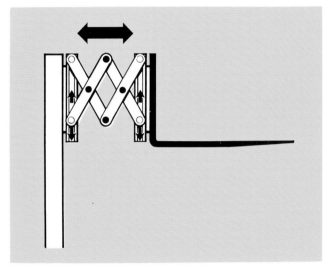

Principle of the pantograph mechanism.

Narrow aisle trucks

Narrow aisle trucks save space and are used primarily for high bay racking operations. They are able to handle pallet loads without the necessity of turning. The required aisle width, generally speaking, is equivalent to the width of the load or truck.

There are three basic types: telescopic fork, rotating fork, and rotating mast trucks.

Telescopic fork narrow aisle truck.

Rotating fork narrow aisle truck.

Rotating mast narrow aisle truck.

Application

Narrow aisle trucks, used mainly for high level pallet operations, are not attached to the racking but are also able to transport pallet loads as well as stack them.

Temporary storage rack locations.

Because rotating fork and mast trucks have to rotate their loads, they require wider aisles than telescopic fork trucks.

Telescopic fork trucks, however, have one drawback: they cannot handle loads at floor level. Because of this they utilize a separate bank of temporary storage racking at the end of the bay which is served by other types of trucks (see picture).

Telescopic fork trucks can handle Euro pallets from the short side only because, otherwise, the forks foul the pallet spacers (see picture).

One way of avoiding the need for intermediate temporary storage facilities is to use telescopic fork trucks which move loads between the pallet racking and, for instance, assembly stations or goods reception.

Rotating fork and mast trucks have the advantage of being able to handle Euro pallets at floor level from the long and short sides.

Maximum lifting capacity of narrow aisle trucks is normally 1,000 kg. Maximum lift height is approximately 12 m.

The main alternatives to narrow aisle trucks for high level stacking are stacker cranes or reach trucks.

Because **stacker cranes** run on rails in the aisles, it is complicated to have them operate in more than one aisle. As a result they are employed in lengthy aisles. Aisle width requirements are roughly the same as for narrow aisle trucks. The advantage of stacker cranes is that they are easily automated. On a manually controlled crane, the operator moves with the crane which means he remains close to the load and has a good view of operations. Stacker cranes often cost more than narrow aisle trucks.

Reach trucks need wider aisles than narrow aisle trucks. At similar stacking heights, they represent a 35% reduction in storage area utilization. As narrow aisle trucks are often able to achieve higher lifts (approximately 9–12 metres as against 6–7 metres for reach trucks) the difference is sometimes even greater. One reason for choosing a reach truck is that throughput may be too small to keep a narrow aisle truck occupied and other uses may have to be devised. Another reason is that the warehouse may also incorporate low level picking for which narrow aisle trucks are less suitable.

The use of narrow aisle trucks is likely to increase. Since they occupy little space and their prices are comparable with those of reach trucks, they are suitable, say, for storage associated with assembly and production. The warehouse may then be divided into several smaller areas located close to the goods' ultimate destination. A narrow aisle truck can then be used for both storage/retrieval and transport.

As narrow aisle trucks can be easily moved from one aisle to another, it is possible to achieve high density storage even at low racking heights. It is reasonable to assume that narrow aisle trucks will be subject to some degree of future automation. Driverless narrow aisle trucks may then, in some cases, compete with automatic stacker cranes.

Design

On a rotating fork truck, there is an articulated mounting on the fork carriage. The carriage moves laterally relative to the truck. To take a pallet from the right, the following procedure is necessary:

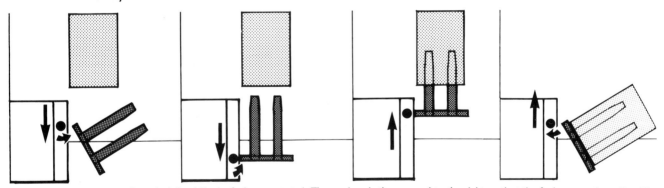

The fork carriage is moved to the left while the forks are rotated. The carriage is then moved to the right so that the forks enter the pallet. The pallet is lifted and the carriage is moved again to the left.

With rotating mast trucks, the actions are similar except that the entire mast is rotated and then moved sideways.

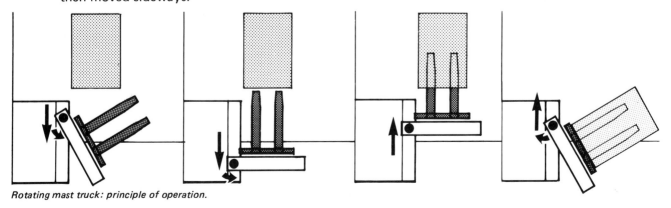

Rotating mast truck: principle of operation.

Rotating fork and mast trucks are often built on chassis similar to those of counterbalanced trucks. They are battery powered and can be equipped with masts of different types. Sometimes, the driver's seat is mounted on the chassis; in other instances he is elevated with the forks. (See 'Order pickers' section).

On telescopic fork trucks loads are handled by means of transversely mounted forks which are extendable from the carriage.

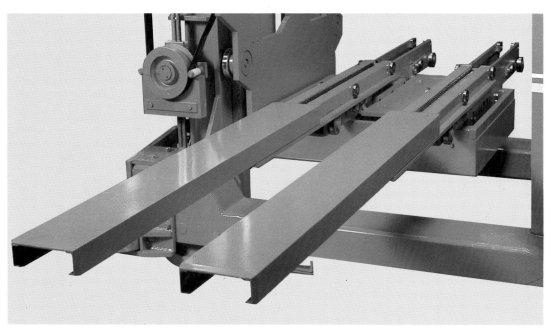

Telescopic forks.

Telescopic forks normally cannot be tilted and only operate horizontally.

Narrow aisle trucks are often fitted with guide wheels running against tracks attached to the bases of both sets of adjacent racking. Normally, four wheels are used for such guiding. Trucks with rigidly mounted masts frequently have two wheels at the top of the mast to increase high level lift stability.

Because the trucks are guided in their aisles, drivers can concentrate on the lifting operations and on driving the trucks forwards and backwards. It is also possible to drive the trucks freely without guide tracks and to change aisles.

Narrow aisle trucks are designed for high level lifting. With telescopic fork trucks, the driver faces sideways which could be a source of strain during lifting operations even though facing sideways is the better option when travelling. Neither a forward nor a sideways facing position is entirely perfect, however.

Dazzle from overhead lights together with distances between driver and load may hinder visibility when using high lift stackers and narrow aisle trucks. Various types of non-glare lighting can improve visibility, while television cameras have been mounted at the forks to give a closer view of things. Experience seems to indicate that it may be difficult to control lift operations precisely with TV cameras. Where possible, it is preferable to do without. Among other aids to facilitate storage and retrieval are height indicating counters, horizontal positioning devices and automatic stacking systems.

Telescopic fork truck fitted with guide wheels which run against tracks mounted on the bases of both sets of adjacent racking.

Order pickers

Order picking refers to retrieval of individual items from storage and normally involves taking goods from one unit load and adding it to another. There are three basic methods.

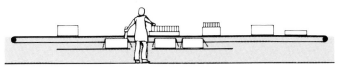

Station picking. The goods come to the operative, for instance, on a roller conveyor. The work station is fixed.

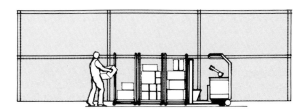

Low level picking. The operative works up to a height of 2.5 metres.

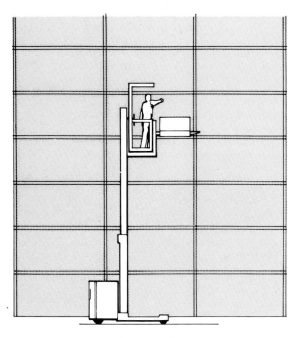

High level picking. The operative is elevated, for instance, with the truck and can thus reach higher storage levels.

Low and high level order picking may involve the use of a number of different truck types.

Low level picking

High level picking

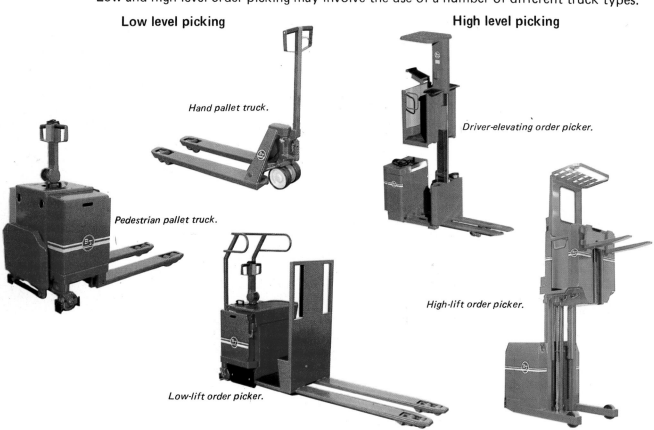

Hand pallet truck.

Pedestrian pallet truck.

Low-lift order picker.

Driver-elevating order picker.

High-lift order picker.

Powered pallet truck.

Hand pallet trucks, the simplest aid to low level order picking, are described in greater detail in the section 'Hand pallet trucks and stackers'. A loaded hand pallet truck is heavy to pull and should not be used for distances over 30 metres or for more than 30–40 pallet movements a day. The weight pulled on a hand pallet truck when picking should not exceed 200–300 kg.

A powered pallet truck relieves the operative from pulling the load.

Pedestrian pallet truck.

Pedestrian guided powered pallet trucks are used for low level picking at higher rates and commonly handle Euro pallets or roll pallets. With extended forks, they can carry two Euro pallets or three roll pallets although such trucks are long and difficult to manoeuvre in some situations. The design of powered pallet trucks is dealt with in the section 'Powered stackers'.

Low level order picker.

The performance characteristics of low level order pickers are basically similar to those of powered pedestrian pallet trucks. The difference is that the order picker is modified for its task. The driver stands on a spacious ride-on platform which is easily boarded. So that the operative can reach higher, a skidproof step-on platform is provided on top of the battery compartment. This is also easy to climb on to. On this upper platform, the driver can pick goods at levels up to 2.5 metres. Low level order pickers have the disadvantage of being longer than pedestrian pallet trucks and thus difficult to manoeuvre in cramped surroundings.

Low level order pickers and pedestrian pallet trucks with extended forks have high capacities and good travel speeds. They are therefore suitable for warehouses with high rates of turnover and for bulky goods.

Where there are many product lines and minimal buffer stock requirements for each line, high level picking systems are appropriate. There are two main types.

Lifting and lowering speeds are rather slow on **driver elevating order pickers**. As the driver has to be lowered to the load carrier after nearly every picking operation, his picking rate too is fairly slow. Such trucks are not normally used to pick at heights above the second or third level of pallet racking and then only for very low picking rates. Where controls can be operated from the driver's platform, efficiency is improved, but the driver still often has to bring the goods down.

Driver elevating order picker lifts the driver only.

For medium and high order picking rates at high levels, a **high lift order picker** is suitable. This lifts both driver and load, is controlled from the platform and can move between different picking locations while driver and load are still elevated.

High lift order pickers move freely in the aisles and their maximum picking height is approximately 7 metres. They can also be guided by tracks for greater stability and for somewhat greater picking heights. Guide tracks also help the driver to concentrate fully on picking. The trucks utilize space efficiently and require no more room than stacker cranes (see section 'Stacker cranes').

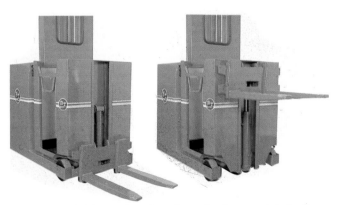

Load carrier height can be adjusted to a level appropriate to the driver's position.

The difference between high lift order pickers and stacker cranes is that, for reasons of stability, the order picker cannot move as fast between different picking locations when the driver is in an elevated position. On the other hand, order pickers can easily be moved from one aisle to another.

Jerky horizontal or vertical movements are unpleasant for the driver, and smooth operation is essential.

High lift order pickers elevate both the driver and the load.

Many high lift order pickers feature **automatic pre-selection of height**. Several picking levels are programmed and the driver need then only press a button to initiate a sequence of operations.

Picking is arduous work and picking operatives must work in positions which are ergonomically efficient. Where pallets are stacked with their short sides facing the aisle, it is difficult to reach the innermost items on a pallet. Long side stacking is preferable although this requires more space.

Initial lift enables the load carrier (on which the driver places the picked goods) to be set at the best height relative to the driver's seat. This allows the operative to take up a good working position whatever load carrier is used.

High lift order pickers capable of free movement must have sufficient travel stability. Maximum driving speed and steer wheel lock, therefore, are restricted when trucks are driven with the cab elevated.

Sideloaders and four-way trucks

Most truck types have difficulties handling lengthy cargo. Piping and steel rods are often 8–9 m long and aisles would have to be absurdly wide if, say, counterbalanced trucks were used. **Sideloaders** and **four-way trucks** have been developed to tackle such work. Both handle loads at right-angles to the driving direction.

Sideloader.

Four-way truck.

The sideloader mast can be shifted in the same way as the mast on the conventional reach truck. Loads can be withdrawn and positioned on the truck's load platform during travel. Diesel engines, cabs and pneumatic wheels are normal features and speeds may consequently be rather high. The chassis resembles that of a lorry with similar transmission and steering arrangements.

Sideloaders can be used both indoors and out of doors (within the limitations governing diesel engines). Aisle requirements correspond to truck widths, with a margin for travel. A narrow truck format allows that the aisles may also be narrow.

The driver sits facing forward and often has to twist his body when picking up and depositing loads and when reversing in the aisles. This occurs especially when the aisles are too narrow for trucks to turn round in them. A narrow cab is, therefore, particularly disadvantageous. Instead, it should be wide enough to provide space for a swivelling seat. Alternatively the cab may swivel; this improves the working position for load handling (see picture).

Four-way trucks are basically reach trucks modified to cope with handling lengthy loads. All wheels swivel and the trucks can move in any direction. They are also fitted with fork-spreaders enabling them to accommodate both lengthy loads and pallet loads. The driver sits facing the load in order to have a good working position for both travelling and load handling. The four-way truck features are described in greater detail in the section 'Reach trucks'.

Sideloader with swivelling cab.

Tow tractors and platform trucks

Tow tractors and platform trucks are used for longer transport distances. Driving speeds are high but some method of loading and unloading the truck or trolley is required. Lightweight materials can be handled manually; heavier goods will require the use of a forklift.

Tow tractor.

In comparison with platform trucks, tow tractors have the advantage that they may be uncoupled from their trolleys and do not, therefore, have to wait around during loading or unloading.

Platform trucks may be fitted with fixed or tilting platforms. They may be battery powered for indoor use or diesel powered for outdoor use. Applications include distribution of loads within a company, transporting sand or refuse, and moving goods which are unsuitable for placing on pallets.

Tow tractors can pull one or several trolleys or carts and thus offer high transport capacity. They, too, may be powered by battery or diesel engine.

Quite often counterbalanced trucks are used as tow tractors. They first load the trolleys, pull them to the reception area and then unload the goods, thus increasing their utilization rate.

Straddle carriers and mobile lifting frames

Straddle carriers and mobile lifting frames are examples of specialized trucks for large loads such as containers.

Straddle carrier.

Terminal tractor and mobile lifting frame.

A straddle carrier envelops its load from above. If high enough, it can move its load above the tops of a line of similar stacked loads and thus requires virtually no operating aisles. Loads may virtually be block stacked. Straddle carriers are common in container handling, although this application is diminishing. If equipped with lifting bars they are also useful for general lifting.

Mobile lifting frames surround their load horizontally instead of from above. Support arms rise up on wheels to lift the load. The lifting frame may have its own power source or may be towed by a tractor; in the latter case more space is required for manoeuvering.

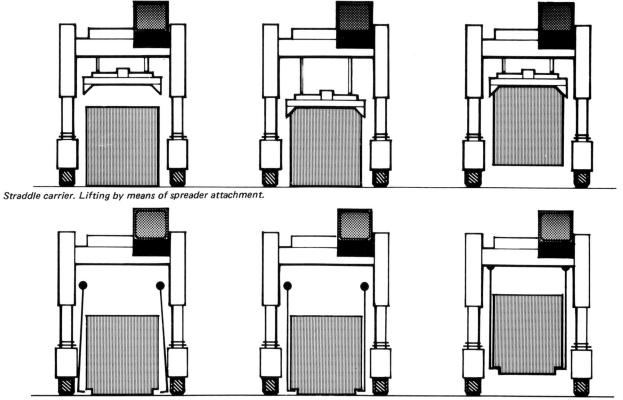

Straddle carrier. Lifting by means of spreader attachment.

Straddle carrier. Lifting by means of lifting bars.

Hand pallet trucks and stackers

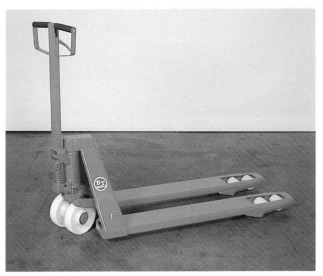

Hand pallet truck.

Loading a lorry using a hand pallet truck.

Hand pallet trucks are the simplest and cheapest pallet handling aid. Wheels are located in each fork tine and there is a steerable wheel at the rear. A pump mechanism in the drawbar provides lift with a hydraulic cylinder elevating the forks relative to the steer wheel; simultaneously, the fork wheels are extended downwards by means of a linkage. The pallet is lifted approximately 0.2 metres. Fork wheel diameter is limited by the depth of the pallet (with Euro pallets, the wheel diameter is approximately 85 mm) and consequently, these trucks require hard and even floors. Starting to move a 1,000 kg load on a hard, even surface requires a tractive force of approximately 15 kgf if the truck runs on nylon wheels. Lifting a similar load requires a pumping effort of approximately 20 kgf. For a load weighing 2,000 kg, these forces are doubled.

Hand pallet trucks may be equipped with hand brakes for use on inclines and to stop a loaded vehicle simply and safely. The hydraulic system can be augmented with a high speed lifting facility which enables 100–500 kg loads to be lifted using one or two pumping strokes.

Hand pallet trucks are used wherever heavy loads have to be moved. Common applications are in terminals, loading and unloading lorries, production and assembly, packing, etc. They may also be towed by powered trucks.

Where loads are large or heavy, a powered pallet truck may be justified. The elevating mechanism is hand operated but travel is by means of a battery powered motor. As a rule of thumb, a powered pallet truck is necessary when 500 kg of materials has to be moved 20 metres 35 times a day.

Powered pallet truck.

Powered pallet trucks are often used where the workforce includes women.

Hand stacker with manually operated hydraulics.

Hand stackers are hand pallet trucks fitted with masts. As handling heavy loads requires much effort, these stackers are used for relatively infrequent lifting operations. Applications include serving lift tables, single lifts of pallets into and from pallet racking (5—10 times a day at the most), etc. Hand stackers may obviously be used in the same way as hand pallet trucks.

As soon as the throughput increases, some type of powered vehicle becomes appropriate. Powered lifting on a hand stacker permits a much greater number of lifting operations.

A powered pallet truck should replace a hand pallet truck when a work shift demands more than 30—40 pallet transfers, when distances are longer than 30 metres or where the drag coefficient is high. A powered stacker should replace a hand stacker (with or without powered hydraulics) when transfer requirements increase.

Hand stackers may be used as work benches.

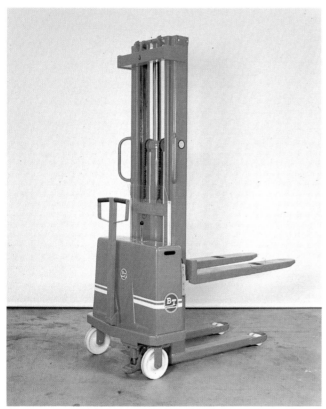

Hand stacker with powered hydraulics.

Hand stacker used in a retail warehouse.

Carts and trolleys

This type of equipment is common for intensive short distance transportation of bulk goods or general cargo consisting of small mixed items. The equipment may be divided into:

- Vehicles with four or more wheels (hand platform trucks, trolleys and trailers) and one or two wheels (hand trucks or carts).
- Lifting and stacking vehicles, i.e., manually moved vehicles equipped with manual or powered lifting (see section 'Hand pallet trucks and stackers').

Hand platform trucks and trailers

The simplest type has a tubular steel frame and a platform made of wood, plastic or metal as well as single or double upright ends. One pair of wheels are fixed, the others swivel freely. Wheels are made of rubber or plastic, diameters are 100–200 mm and bearings are either needle or ball races.

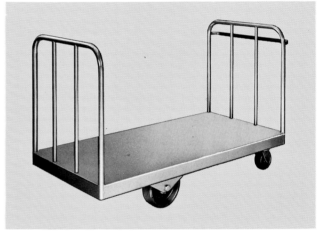

Hand platform truck with two fixed upright ends.

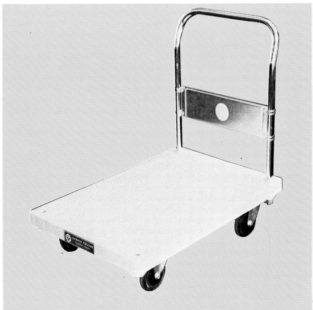

Hand platform truck.

On such a **hand platform truck,** the 700 × 1,000 mm platform may be 180 mm above floor level and the capacity may be 200–500 kg. To save storage space, the uprights may be collapsible. If sides are added the platform truck becomes a **box truck.** The sides may be of wood or metal mesh; sometimes there is a door. If the vehicle has a basic plan of 700 × 800 mm, it becomes a **roll container** or **roll pallet** suitable, for instance, for retail trade deliveries (see section 'Load carriers').

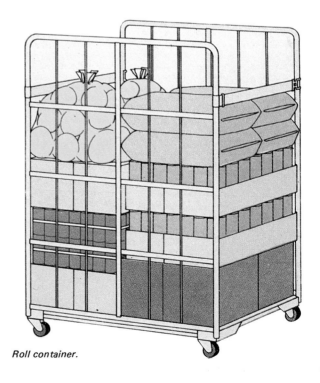

Roll container.

92

Small wheels (to reduce platform height) and a rigidly mounted handle result in a trolley that resembles a shovel for loads up to 200 kg. Normally there are two fixed wheels and a castor beneath the handle junction.

One version of the hand platform truck is built to move bins of standardized dimensions and is common in self-service and department stores.

By increasing the floor clearance and the overall height, the carrying surface may be raised to match table level. Additional shelves increase the carrying area. Usually all four wheels are castors and the carrying surface may be plastic, glass or sheet metal. Where hygiene is a factor, stainless steel is used.

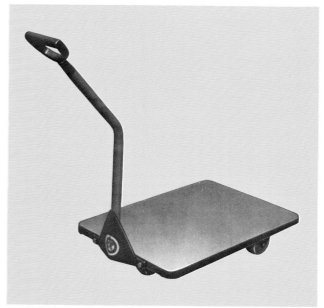

Low loading trolley with rigidly mounted handle.

Order picking trucks for use in warehouses normally have a single fixed pair of wheels. An additional handle, a picking list holder and sometimes a small stepladder are also included. Special purpose fitments include a roller bed or cantilevered racking for lengthy loads.

Offering a larger load area and larger, pneumatic wheels, the **platform trailer** is often supplied with sides and ends or with corner supports. The drawbar is connected to the steerable front wheels which are linked by a tie rod. Wheel diameter is 350–400 mm, the platform area may be up to 1,000 × 2,500 mm and the load capacity 1,000–1,500 kg. The rear wheel disc brakes may be locked by means of a lever or pedal. The drawbar is normally designed both for hand towing and for hitching to a tractor.

Table trolley.

Platform trailer.

Order picking truck.

The timber trailer has an open frame and four corner posts. It is typically 1,100 mm wide and has a wheelbase of 2,000—3,000 mm and a 2—2.5 tonne load capacity. It may be pulled manually or by a vehicle. Turning radius is 2.3—3 metres.

Timber trailer.

The wheel configuration for the above trailers is usually four load carrying wheels, two of which are castors. However, there are alternatives:

- Two wheels at the centre of the trailer plus one front and one rear. The latter are mounted so that only one of them at any time supports the trailer and its load. By seesawing the trailer the front and rear wheel may be used alternately, thereby improving the turning radius.

- Two wheels at the centre with pairs of wheels front and rear. This system functions similarly to that above but permits the trailer to accommodate heavier loads.

According to US tests, trailer wheel drag is acceptable if the tractive force needed to keep the trailer in motion is 120—140 N (12—14 kgf), with an additional 100—150 N (10—15 kgf) required when starting. Experience shows that, to reduce the feeling of drag, the trailer should be pushed rather than pulled. Rolling conditions can often be improved by using harder wheels on soft surfaces, and vice versa.

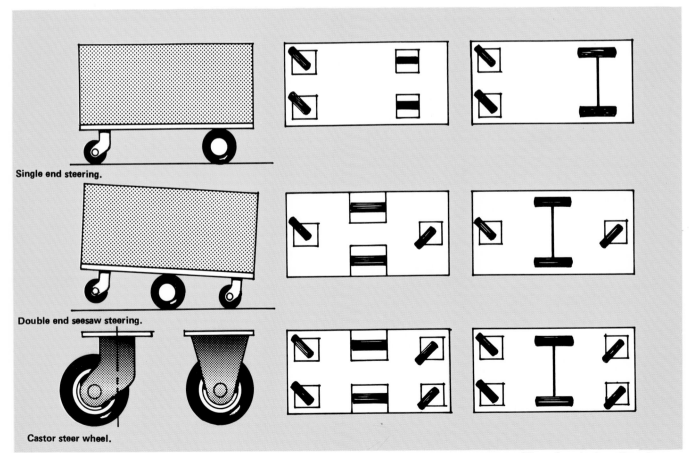

Single end steering.

Double end seesaw steering.

Castor steer wheel.

Alternative wheel configurations.

Carts

Single wheel barrows and twin hand carts vary in design according to the product carried (e.g., concrete, bricks, cereals). Two wheel sack trucks are inserted under the sack which is then tipped back against the frame during transport. Sack trucks may have a rigid or folding toe and are designed for loads up to 300 kg. They can also be adapted for special loads such as gas cylinders or drums. Cleaners' trucks have bicycle wheels and refuse sack holders, builders' trucks handle bags of cement and stacks of bricks, stair crawlers negotiate stairs. The latter may have a rotating set of wheels or a powered caterpillar track. A roller crowbar may be used to shift heavy loads in confined spaces.

Sack truck.

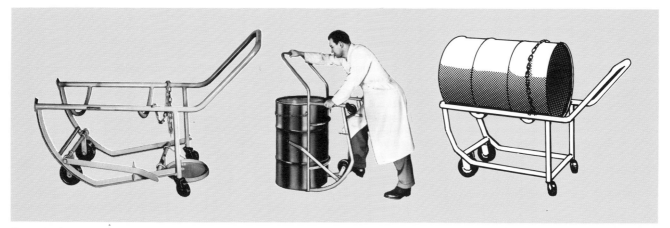

Drum truck.

Stair crawler.

Elevating appliance truck. The pair of wheels can be raised or lowered on the mast for loading and unloading lorries on locations where there are no loading bays.

A powered stair crawler for loads of up to several hundred kg incorporates a battery and motor which drive a ribbed rubber track to propel the truck upstairs. The truck is comparatively heavy (60 kg approximately) and less adaptable on the level than a simple sack truck.

Bin handling carts and trolleys

Special vehicles have been developed for handling single or stacked plastic bins. This development should be seen as improving the working environment by reducing heavy manual work to a minimum. Handling operations in abattoirs and other locations associated with the production and distribution of perishable produce are now based on special trucks whereby stacks of plastic bins may be lifted directly from the floor without prior lifting by hand.

This truck grips the purpose designed bins on both short and long sides.

This four wheel truck lifts bin stacks via arms which engage with ridges on the bin sides.

Air-cushion handling frames

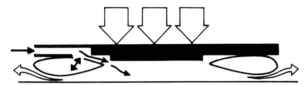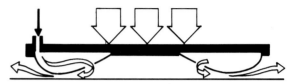

Two different types of intermediate air-cushion device seen in cross-section.

In air-cushion handling devices a rubber diaphragm is bonded or bolted to the load carrying surface. Air, provided by a fixed supply system, is blown continuously through the support structure into the diaphragm and out through holes in the bottom or the top. The build-up in pressure creates a lifting force which, when it exceeds the weight of the load carried, causes the air-cushion to become airborne, allowing air to leak out between diaphragm and floor. This leakage through a 0.2 mm gap reduces the coefficient of friction between the membrane and the floor to 0.001–0.003, and the air-cushion may be moved in any direction.

Air-cushions are used, for instance, to move machinery without the need for separate load carriers.

In the most common applications however, air-cushion units are mounted beneath some form of load carrier.

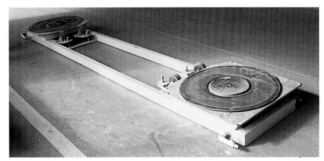

Underside of a frame fitted with two fixed air-cushions. Large objects may require two or more frames.

Air-cushion characteristics:

Advantages	Disadvantages
Low floor surface pressure (5–30 N/cm²). Low tractive force required, even with heavy objects. Movement in any direction. Exact load positioning and vibrationless motion. Accident risks slight. Moves bulky, heavy objects. Small capital cost, especially relative to operations involving heavy loads. Very low operating and maintenance costs. Permits flexible production layouts. Frees floor areas for other uses.	Requires good quality floors. Low transport speeds. Air supply source required. Only horizontal load movements.

Air-cushions require good quality floors:
- Class 2 evenness according to Hus-Ama.
- Surfaces capable of giving an air seal.
- Cracks ≤ 0.3 mm wide.
- Height differences at joints ≤ ± 0.5 mm.
- Surface irregularities ≤ 0.2 mm.

Costs of laying floor topping (1980)	
Material	Cost SEK/m²
Mastic asphalt	55
Granolithic concrete	70
Epoxy coating of concrete floors	25

Application examples

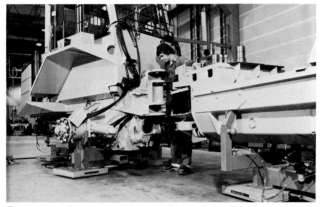

Forestry equipment assembly station: air-cushions mounted on fixtures.

Excavator assembly line in which the chassis rest on air-cushion platforms.

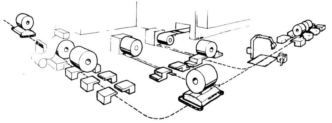

Automatically guided air-cushioned vehicles for moving coils of sheet metal in a rolling mill. Guidance is via wheels which carry a fraction of the total weight.

Examples of air-cushion specifications						
Capacity kgf	Diameter mm	Height mm	Weight kg		Air pressure kgf/cm² atm. gauge	Capacity m³/min free air at max. load
			To be mounted	With load support		
200	200;250	22;29	0.7;1	2.7	0.6;0.9	0.1;0.2
500	310;360	22;29	1;1.4	3.2	0.7;0.9	0.2;0.3
700	430	22	2.3		0.7	0.2
1,000	310;510	22;29	1.5;2.7	4.7	0.8;1.8	0.3;0.4
1,600	400;600	22;29	2;4	4.5	0.7;1.8	0.4
2,300	760	25	5.9		0.6	0.6
2,700	920	32	8.2		0.5	0.7
3,200	530	51	3	10	1.8	0.4
3,600	1090	32	11.8		0.5	0.9
5,400	690	62	7	18.6	1.8	0.4
6,400	530	51	4.5	10.9	3.5	0.6
10,900	690	62	9	23.6	3.5	0.6
18,200	910	68	23	44	3.5	0.8
36,300	1220	68	45	71.7	3.5	1.2

Platforms should be supported by not less than four air-cushions. Platform sizes vary with user requirements. Handling frames require at least two air-cushions. The table below shows the prices of a number of air-cushion units, based on suppliers' average prices (1980).

Examples of air-cushion equipment prices				
Capacity kgf	Capital investments exclusive of VAT, SEK/unit (1980)			
	Air-cushion modules		Platforms	Frames
	To be mounted	With load support		
2,000	900	1,800	7,000	3,000
5,000	1,600	3,500	10,000	5,500
10,000	2,800	6,200	15,000	9,500
20,000	5,300	11,600	24,000	17,000

Overhead travelling cranes

An overhead travelling crane consists essentially of a bridge constructed from one or several girders supporting a mobile hoist. End carriages drive the bridge along rails on two high level gantries.

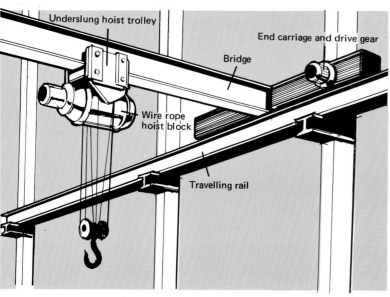

Indoors, the rails may be mounted on pillars or suspended from roof supports.

Pillar supported, rail mounted overhead travelling crane.

Application

Overhead travelling cranes are used mainly for handling heavy, bulky or lengthy loads such as steel plates, fabrications, sub-assemblies and complete items of machinery. Loads are handled within an area bounded by the crane's span and the length of the travelling rails, leaving the floor area completely free for other activities. As a result, a wide range of applications can be made in, for instance, steelmaking, metal working and fabrication.

Other applications include indoor and outdoor storage installations for lengthy items such as pipes, tubes, girders and sections. The area under the crane is utilized to the maximum while the crane can be used simultaneously for loading and unloading lorries or railway wagons. Large cargo terminals also employ travelling gantry cranes for loading, unloading and the short-term storage of heavy and lengthy loads. In modified form, overhead cranes can be used for storage and retrieval in warehouses.

Overhead travelling cranes, operating indoors, do have one disadvantage in that they cannot usually serve the full volume of the building since the working volume they encompass ends 1—5 metres from the walls and ceiling.

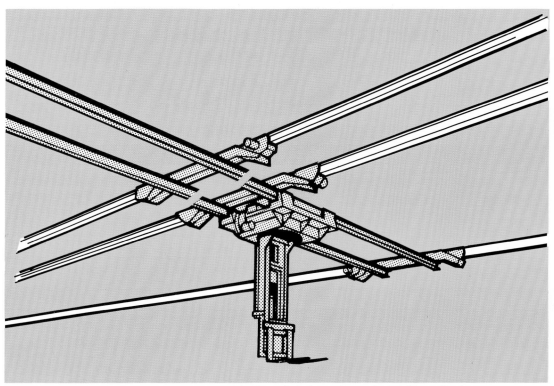

Double girder underslung travelling gantry crane with telescopic fork attachment for handling pallets in a warehouse.

Advantages and disadvantages of overhead travelling cranes	
Advantages	Disadvantages
Ability to handle heavy and bulky goods Floor areas left clear	Reinforced building frames often required Slow operations Risk of 'queueing' Inability to utilize total area of buildings Use restricted to a given area

Design

Overhead travelling cranes are either **top mounted** or **underslung**. In **underslung** designs, the end carriages 'hang' on a roof mounted I-beam. Top mounted cranes run on girders or rails supported by pillars (see pictures).

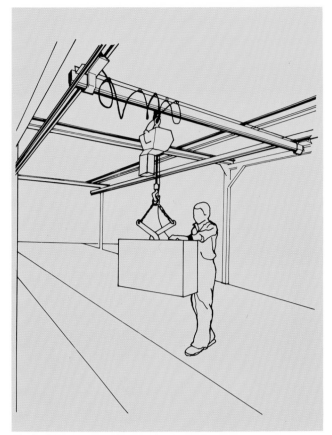

Single girder, top mounted overhead travelling crane.

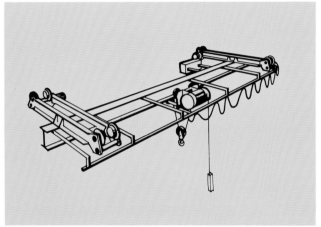

Underslung overhead travelling crane.

For heavier loads **top mounted cranes** are the most commonly used. They come in standard versions with capacities of 2–50 tonnes and in custom-built versions up to 500 tonnes. Single or double girder structures are available, the former being sufficient for loads weighing up to approximately 10 tonnes, the latter being used for heavier loads.

Overhead travelling cranes usually have three main drive mechanisms for:

- Lifting.
- Crane travel.
- Hoist trolley travel.

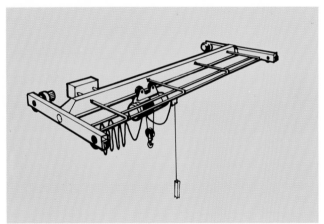

Manual travel overhead crane.

Underslung cranes weigh less and require less headroom than top mounted cranes; they also permit better space utilization. Pillars are not needed and the hoist trolley can utilize the entire span of the crane and even change over to another crane.

However, the lifting capacity is limited by the strength of the roof. As a result, underslung cranes are normally available with capacities of 5 or 6 tonnes, occasionally 10 tonnes.

For lighter loads (up to 1,000 kg) simple crane kits are available. They incorporate chain hoist blocks and, since they are usually not equipped with a motorized trolley, horizontal travel is manual.

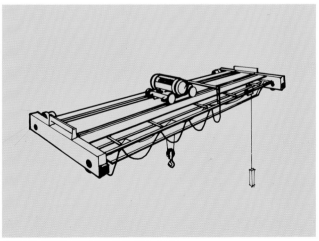

Double girder overhead travelling crane.

Single girder cranes normally feature a wire rope hoist block mounted on an underslung trolley running on the bottom flange of the crane bridge. Double girder crane bridges have hoist trolleys which run along the tops of both beams. The beams are usually in the form of steel box fabrications.

The crane travel mechanism propels the entire assembly on girders or rails. The hoist trolley has its own travel drive.

The efficiency, speed and reliability of overhead travelling cranes depend to a great extent on the parallelism of the travel rails and the synchronicity between the end carriage motors.

The power units for both travel and lifting operations are usually electric motors. Hydraulic and pneumatic motors are also used for lifting and may provide control advantages.

However, electric motors are continually being developed for improved power—to—weight ratios and better control characteristics. Some electric drives, for example, can be reversed without first being brought to a halt, a facility which makes for considerably quicker and more flexible handling.

Standard trolley with wire rope hoist block for single girder crane.

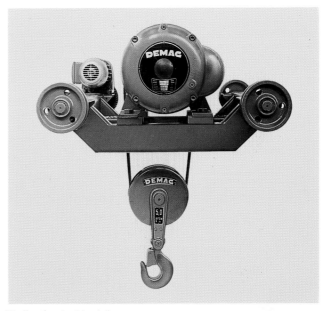

Trolley for double girder crane.

Cranes for lightweight applications are usually operated from floor level by means of a pendant control (see picture). On heavier cranes, for instance, handling steel, cabs (fixed or designed to move with the crane) are more common.

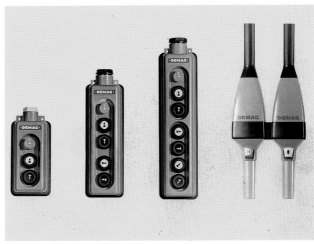

Low headroom trolley with wire rope hoist block.

Pendant controls for lightweight cranes.

System design

When installing an overhead travelling crane, it should be regarded as an integral part of the building. The drawing below indicates various dimensions and load points which may be of importance and which determine crane capacity, reach and lifting range. (The example refers to an underslung crane.)

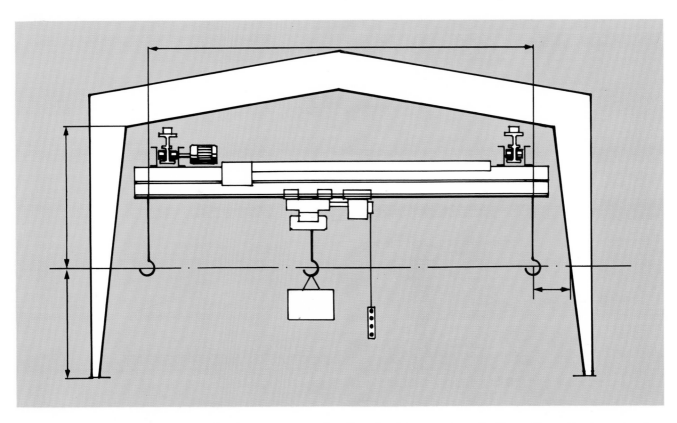

In addition to the usual lifting hook, cranes can be fitted with a range of alternative attachments, for example, tongs, clamps, grippers and magnets (see figures).

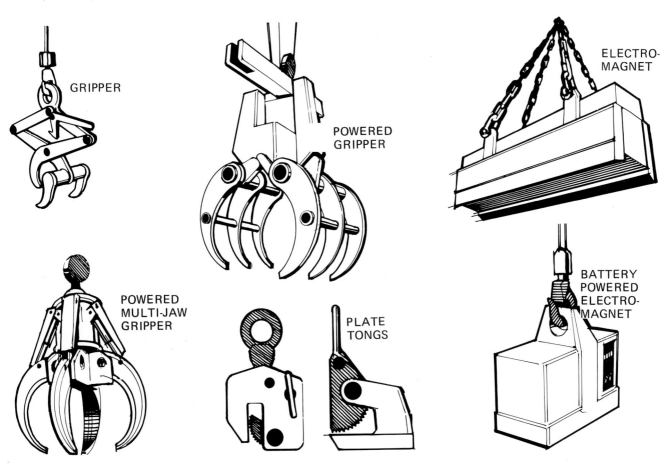

GRIPPER

POWERED GRIPPER

ELECTRO-MAGNET

POWERED MULTI-JAW GRIPPER

PLATE TONGS

BATTERY POWERED ELECTRO-MAGNET

Application examples

The photographs below show examples of overhead travelling crane applications.

Manual overhead travelling crane.

Top mounted overhead travelling crane with telescopic clamp for paper reels.

Underslung double girder overhead travelling crane with cab and extendable loading device for lengthy loads.

Top mounted overhead travelling crane handling piping and tubing. The floor area is well utilized, but not the potential volume.

40 tonne rail mounted double girder overhead travelling crane for foundry work.

EXAMPLES OF OVERHEAD TRAVELLING CRANE SPECIFICATIONS

Crane type	Lifting capacity tonnes	Lift height m	Max. span m	Building height m	Max. hoisting speed m/min	Trolley speed m/min	Crane travel speed m/min
Lightweight under-slung crane fabricated from plate and equipped with chain hoist block	0.25 1.0 1.0	8–16 8–16 10	2–5 2–5 5	1.0 1.0 1.0	5–15 4–8 4.6	12–24 12–24 18	– – 20
Single girder under-slung crane	1.0 5.0 5	10 4–10 10	12–18 12–18 15	1.0 1.3–1.5 1.5	8 4–8 8	10–20 15	15–30 15–30 20
Single girder top mounted crane with hoist block	2.0 10.0 10	10 10 10	10–20 7 7	1.0 1.5 1.5	8–10 6–8 6.3	10–20 10–20 15	17–50 17–50 30
Double girder top mounted crane with hoist trolley	16.0 50.0 32	12–16 10–20 10	25 36 30	1.3	10 2–6 4	20–30 25–60 40	25–60 25–120 50

Gantry cranes

Gantry cranes (or Goliath cranes) are supported on legs with wheels that run on rails or on the ground. A hoist trolley traverses the main girder. The hoist mechanism may be installed in the trolley or inside the gantry structure.

Gantry crane for container handling.

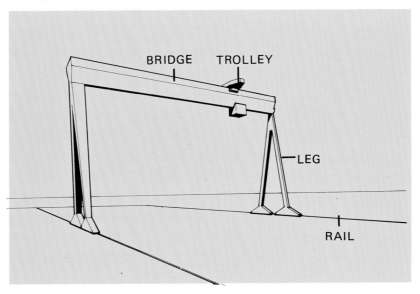

Application

Gantry cranes are used primarily for very heavy loads (over 20 tonnes). A rail mounted gantry crane is basically an overhead travelling crane for outdoor work and is used for the same type of jobs — transhipment, unloading, stacking and storing awkward and heavy loads such as steel plate and containers, and moving bulk materials, e.g., scrap. Outdoors, a gantry crane is, generally speaking, a less costly solution than an overhead travelling crane, as a fixed support structure is not necessary and the gantry legs replace a long line of pillars. On the other hand, a gantry crane is usually slower in operation. Because the moving mass is greater and the structure less rigid, it is difficult to achieve comparable travel characteristics at a reasonable cost. If the speed and track length requirements are only moderate, therefore, gantry cranes may be an alternative for outdoor work.

Gantry cranes appear in many sizes, from 20 tonne container cranes to shipyard units capable of lifting entire prefabricated ship sections weighing up to 1,500 tonnes.

600 tonne gantry crane for handling entire ship sections.

Advantages and disadvantages of gantry cranes	
Advantages	Disadvantages
Can cope with long, heavy loads over lengthy transfer distances	Restricted view for operator could pose safety hazards
Cheaper than overhead travelling cranes	Slower than overhead travelling cranes
Can cover large areas	

Design

Gantry cranes travel on rails (for heavy applications over 50 tonnes) or may be fitted with pneumatic tyres. Normally there are four legs mounted on bogies with a suspension system which distributes the load equally between the wheels.

Large cranes demand great structural rigidity and synchronized travel motors. To allow for variations in track gauge and crane span due to thermal expansion, one pair of legs is pivoted.

A distinction is made between full gantry and semi-gantry cranes. The latter has one end which travels in the same manner as an overhead travelling crane. Semi-gantry cranes are mainly for indoor use, often complementing an overhead travelling crane.

The gantry is either a box girder or a lattice girder structure. The latter type can come in a modular kit form enabling cranes to be purpose-built from standard components.

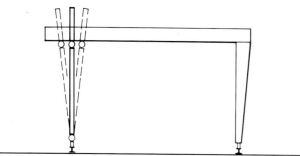

Gantry crane with pivoting leg.

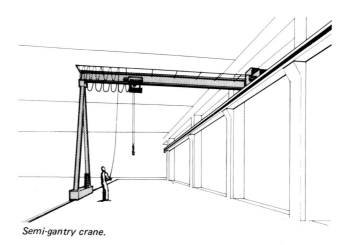

Semi-gantry crane.

Lattice girder gantry crane.

Hoist trolleys can be straddling or underslung. Using both these techniques on large gantry cranes, allows trolleys to pass each other (see figure).

A lattice structure also allows the hoist trolley to run inside the structure.

The hoist is often mounted directly on the trolley, at least in the case of heavier capacity cranes. For gantry cranes in the 20–30 tonne category, there is a simpler arrangement: two cable drums mounted on the crane transmit power for both trolley traverse and hoisting.

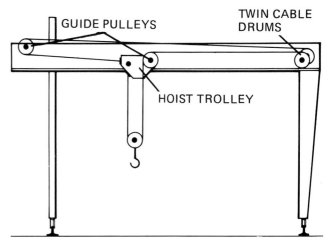

GUIDE PULLEYS

TWIN CABLE DRUMS

HOIST TROLLEY

Gantry crane with fixed mechanism for powering hoisting and trolley traverse.

The hoists are basically similar to those used with overhead travelling cranes (see section 'Overhead travelling cranes').

STRADDLE TROLLEY

CRANE BRIDGE

UNDER-SLUNG TROLLEY

Gantry crane bridge with two hoist trolleys.

Application examples

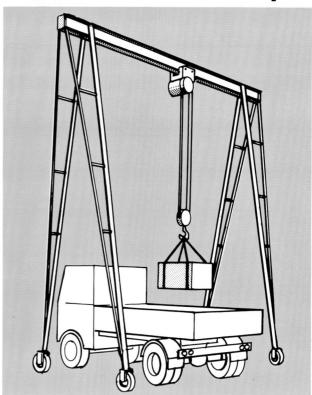

A lightweight gantry crane equipped with a hoist block unloading heavy, awkward goods from a lorry. An appropriate solution where there is a need for simple, versatile equipment capable of infrequent handling of goods which can be stored out of doors.

Gantry cranes have become a standard method of moving containers between railway wagons and lorries. At SJ container terminals these 30 tonne rail mounted cranes with straddle hoist trolleys and spreader attachments, cope with containers and platforms of various sizes.

Container ports use gantry cranes not only for loading and unloading ships but also for internal handling. Rail mounted and rubber tyre versions are employed and, in both cases, crane movement may be optimized electronically. Cranes on rubber tyres are then controlled by means of underground induction loops.

Gantry crane on rubber tyres for container stacking.

Jib gantry crane for loading and unloading containers.

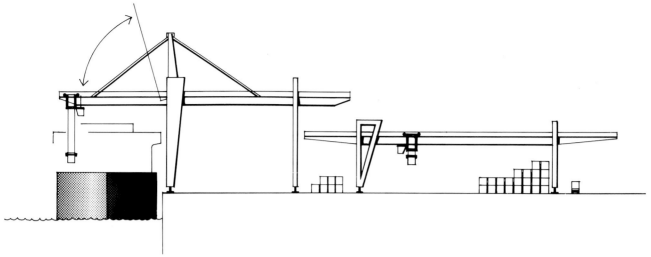

Container ports may use a gantry crane combination in which one crane loads and unloads ships while the other stacks and sorts as well as loads road vehicles and railway wagons.

A steel stockholder uses a 12 tonne gantry crane controlled from the cab and equipped with straddle hoist trolley and magnetic spreader attachment. The crane straddles the entire storage area (60 m wide) and provides maximum area utilization as no aisles are needed. Travelling speed is 63 m/min.

A lightweight gantry crane on wheels moving general cargo. The winch position is fixed and there are traversing hoist blocks for each pair of gantry legs. Because the structure flexes, the crane is capable of travelling safely over uneven surfaces.

Summary

Broadly speaking, gantry cranes and overhead travelling cranes have similar specifications.

On the whole, gantries cost slightly less than overhead cranes which operate from fixed support systems.

Stacker cranes

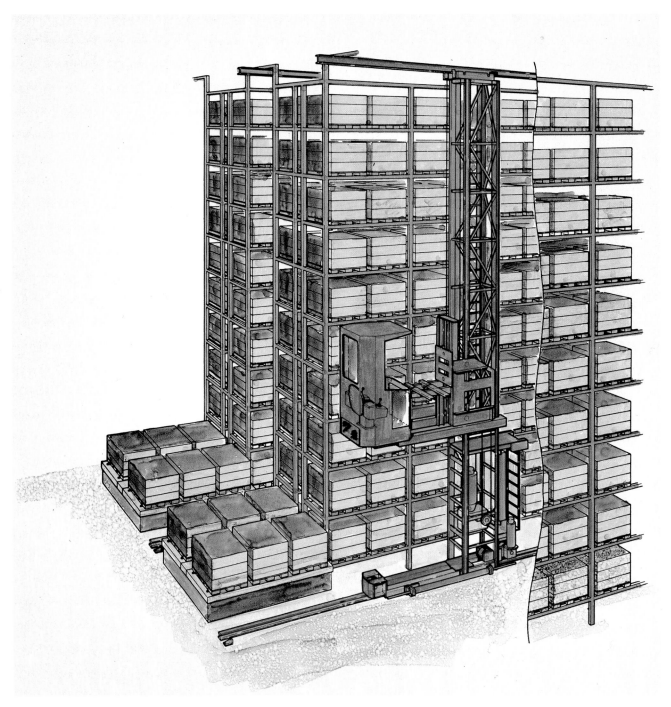

Stacker cranes are rail-mounted devices providing storage and retrieval in high bay warehouses, i.e., racking installations 10–35 m high which would generally be beyond the scope of forklift trucks, reach trucks or narrow aisle trucks. Since stacker cranes can operate in narrower aisles than conventional lift trucks, they can increase the storage potential of a given floor area by 3 to 5 times, though this improvement would be smaller if the comparison was with narrow aisle trucks. In addition to *savings in volume* there are also considerable *labour savings* since stacker cranes offer automation opportunities.

Automated warehouses create further advantages such as improved *materials administration* (i.e., the possibility of reducing inventory and, thereby, tied-up capital), reduced *damage to goods* and *wastage,* simpler *stocktaking,* ability to *'block' quarantined goods,* and *integration* with production and distribution as well as with administrative systems.

Crane driven storage systems are expensive: a stacker crane will cost from SEK 200,000 to 500,000 (1980) depending on its lifting capacity, control system, etc.

Stacker and order picker cranes are used in the following types of operation:

- Manual or automated storage and retrieval of *entire pallet loads.*
- *Manual order picking* from pallet racking and shelving.
- In *specialized racking* designed for manual or automatic storage and retrieval of other *unit loads.* Examples are live storage pallet handling, cassette storage of lengthy loads on cantilevered

racking, handling 20 ft ISO containers to and from container stillages and special load carriers bearing car bodies, lorry cabs, cars, etc.

In Sweden there are at present (1980) 150 installations with 360 stacker or order picker cranes.

Stacker cranes are only one component of storage installations which also incorporate:

- Racking.
- Systems for feeding goods to and from the cranes (conveyors, automatically guided vehicles, driver operated trucks).
- Control systems.
- Buildings which include electrical supplies, ventilation, heating and sanitation, fire protection and staff premises. Walls and roofs usually consisting of some form of insulation attached to sheet metal may be fitted straight onto the racking instead of housing the racking within a conventional steel and concrete structure. Cold store variants are also possible.

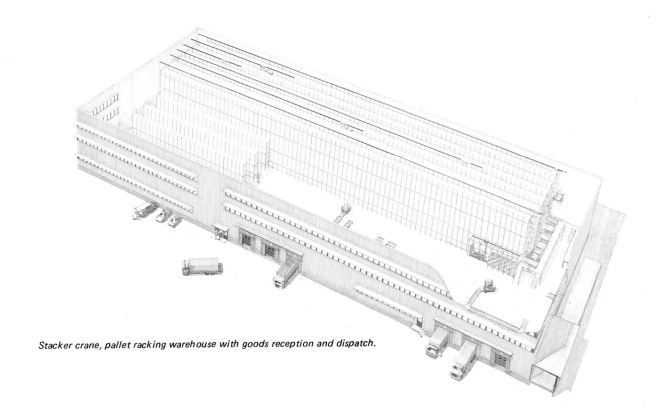

Stacker crane, pallet racking warehouse with goods reception and dispatch.

Design and performance

Definitions
The following definitions occur in the Swedish International Safety Board Directions No. 34, 'Stacker cranes':

Stacker crane: Lifting and transport device laterally guided by rail or guide tracks, for stacking pallets in pallet racking or for manual order picking of goods from shelves or pallet racking. Control may be manual or automatic.

Order picking stacker crane: Stacker crane used mainly for manual order picking from shelves.

Transfer trolley: Device used for transferring stacker crane from one aisle to another.

Aisle: Space between pallet racking and/or shelves where a stacker crane operates.

Load carrier: Forks, order picking tables, etc., which carry pallet loads or individual items.

Positioning system: Equipment for positioning stacker cranes precisely for storage and retrieval.

Load platform: Vertically travelling carriage which supports load carriers, loads and often an operator's cab.

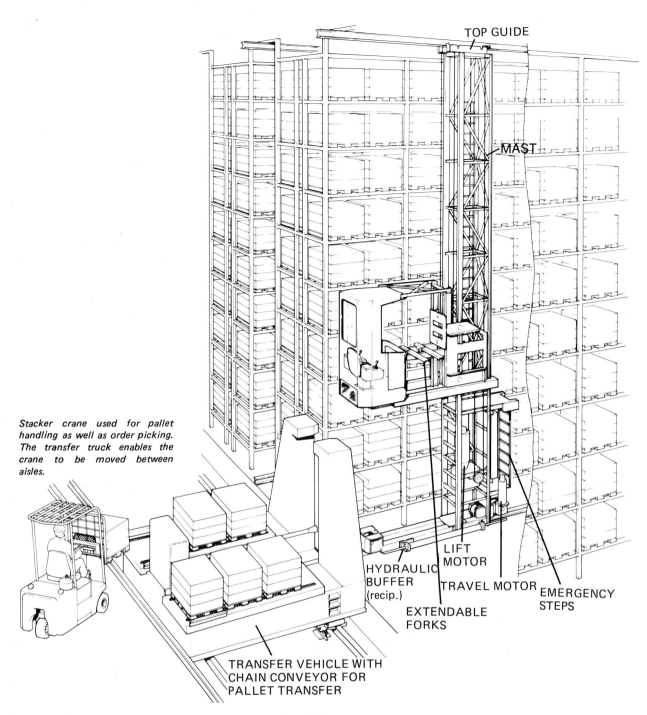

TOP GUIDE

MAST

Stacker crane used for pallet handling as well as order picking. The transfer truck enables the crane to be moved between aisles.

LIFT MOTOR

HYDRAULIC BUFFER (recip.)

TRAVEL MOTOR

EMERGENCY STEPS

EXTENDABLE FORKS

TRANSFER VEHICLE WITH CHAIN CONVEYOR FOR PALLET TRANSFER

Aisle-based stacker cranes were developed in the 1960s as a variant of overhead stacking cranes. The latter were mounted on rails on top of the racking. Nowadays cranes normally run on ground-level rails and are guided at the top of the racking only to ensure lateral stability. Another variant was suspended from a point half-way up the racking; this had the advantage of generating less mast vibration when travel speed was changed.

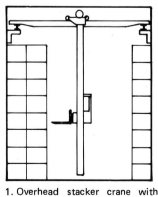

1. Overhead stacker crane with rotating mast.

Development stages of stacker crane.

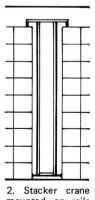

2. Stacker crane mounted on rails on top of pallet racking guide at floor level.

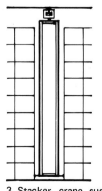

3. Stacker crane suspended from roof beam; guide track at floor level.

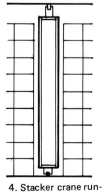

4. Stacker crane running on floor rail; guide track at roof level.

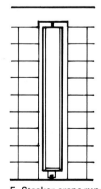

5. Stacker crane running on floor rail; guide track at top of racking.

A stacker crane collects a load at the end of the aisle, moves it to a given location in the racking and deposits it there. During retrieval, the order of operations is reversed. The location is based on co-ordinates defining horizontal and vertical distances and rack number.

The crane itself consists of a mast, load platform (with load carrier), travel and lift devices, cab and control equipment. Single mast cranes are suitable for heights up to 15—20 m; for greater heights and for heavy or bulky loads, twin masts are employed, with the load platform located between them.

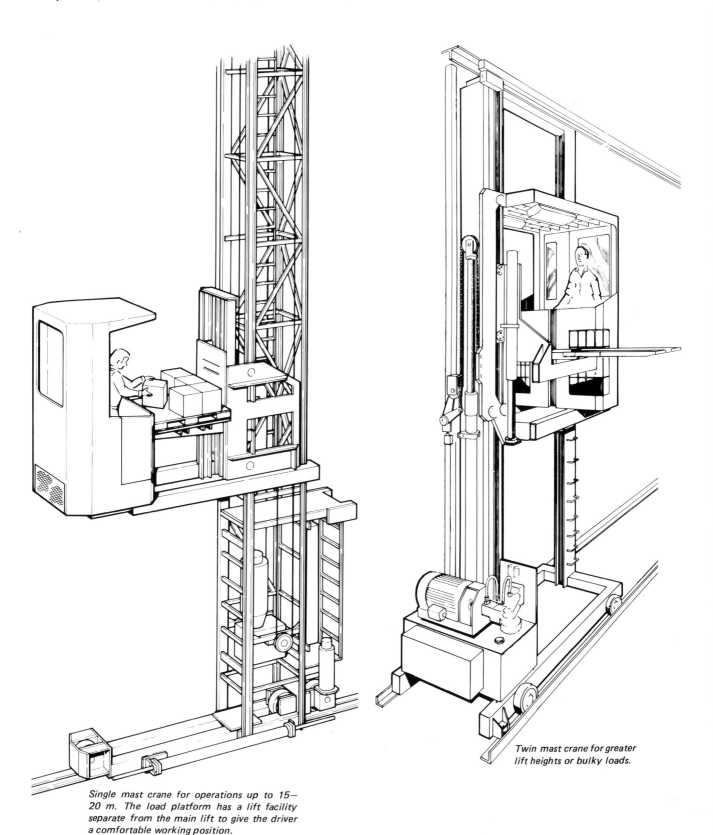

Twin mast crane for greater lift heights or bulky loads.

Single mast crane for operations up to 15— 20 m. The load platform has a lift facility separate from the main lift to give the driver a comfortable working position.

Normally, the load platform supports both the forks and the driver's cab. Single mast cranes have a cab positioned adjacent to the tower or at the extreme end of the platform. The choice of position is influenced by individual ideas on field of vision, operator comfort (vibration), etc.

Fork assemblies are these days designed exclusively as three stage extendable units operated by electric or hydraulic motors in combination with chains and racks. This produces a quick and positive action for storage or retrieval. A complete cycle of operations may take only 12–18 seconds. For live storage operations, the fork assembly is usually replaced by a powered roller bed section; in other live storage designs, the load carrier acts as a shuttle truck (connected by cable or self-contained) parking goods on roller lines in the racking.

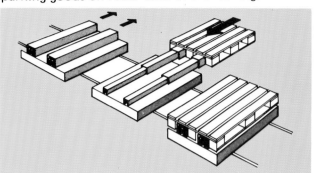

Extendable forks, normally three-stage, electrically or hydraulically operated.

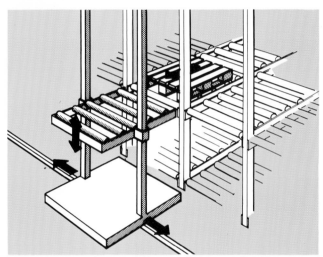

For live storage, the extendable crane forks are replaced by powered roller conveyor.

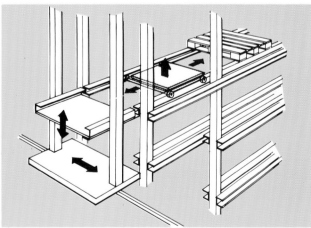

Shuttle truck feeding a pallet into racking.

Fork assemblies have two synchronized forks. According to the shape and size of the load, however, there are also designs with a single, broad fork or with several forks (the number being determined by the size of the individual loads). Lift and travel may be simultaneous if required, and are produced by separate 5–25 kW electric motors.

Typical stacker crane performance figures are:

● Travelling speed 90–180 m/min.

● Lifting speed 10–40 m/min. (The lift-to-travel speed ratio is based on the storage height to storage length relationship, i.e., $V_{lift}/V_{trav} = H/L$. The ratio is usually set at 1/3 or 1/5).

● Acceleration/retardation 0.3–0.5 m/sec².

● The aisle width should have sufficient margin on either side, normally 100–150 mm. Thus, a 1,200 mm pallet handled on the shorter side, requires an aisle width of 1,400–1,500 mm.

● Load capacities for different scales of operation, with upper limits at, say, 600 kg, 1,000 kg and 1,500 kg. Crane deadweights 5–12 tonnes according to crane dimensions.

● Fork thickness 60–65 mm for normal loads. Heavier loads may demand thicker forks, thus necessitating spacers under the pallets to ensure sufficient handling clearance. Spacers are also required where pallets are handled on the long side since extendable forks cannot penetrate the pallet entirely from this side. Note also that the cranes cannot pick up pallets directly from floor level.

● Distance from floor to first storage level 600 mm, approximately (conditional upon the basic

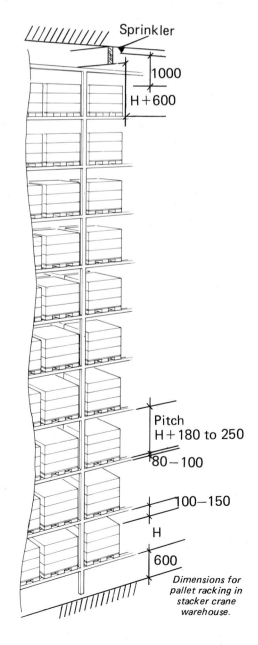

Dimensions for pallet racking in stacker crane warehouse.

crane construction.) The distance from the top storage level in the racking to the roof should be not less than the load height plus 600 mm. The 1,000 mm sprinkler requirement can often be met by positioning the sprinklers between the roof beams.

Power and operation

Stacker cranes are powered by electric motors. Experiments with linear motors have not yet led to commercial availability.

Electric operation normally involves three motors: one for travel (x axis), one for lift (y axis) and one for the forks (z axis). For high levels, there may also be synchronized top drive in order to avoid skew loads and vibration.

For transmitting power between mains supply and cranes, busbars or festoon cables may be used. Compared with festoon cables, busbars are more compace, occupy less space and simplify disconnection when a crane is transferred to another aisle.

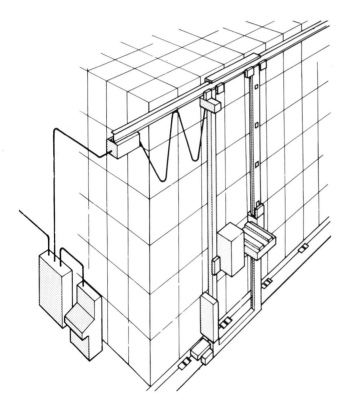

Busbar for electric power transmission to crane.

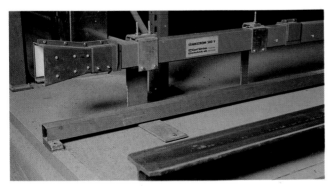

Today busbars have replaced festoon cables. Being enclosed, they may be mounted at floor level.

Crane control comprises three activities: supplying information, information transfer, and execution.

Supplying information means that a two-way flow of decisions is necessary between the control system and the crane in respect of each transaction. Four levels of control may be perceived, namely:

(a) Full manual control, i.e., the driver sits in the

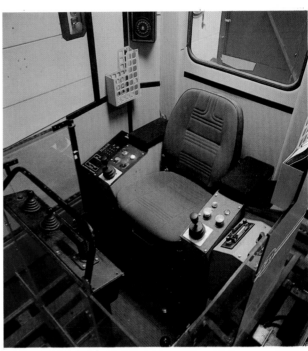

Driver's cab for manually controlled crane. Simultaneous horizontal and vertical travel is possible. A 'dead man's handle' halts operations if the driver leans out.

crane and drives it to its destination, using the controls. The address is obtained manually from a card or computer printout.

(b) Manual control, as in (a) above, but with an address console on the crane enabling the operator to program addresses, using a keyboard, decade switches or punched card reader.

(c) Semi-automatic control with the operator seated at a floor-mounted control centre. Address input as in (c). The operator still enters addresses by hand but is able to handle several cranes simultaneously.

(d) Fully automatic control using a minicomputer to select appropriate rack locations, transmit information to the crane's control system and update continuously the contents of the warehouse (inventory control).

Each variant includes certain administrative routines. When depositing goods, the storage location must always be determined on the basis of load size. At the same time, data are recorded on the product line number, quantity of items per load unit and possibly the date of deposit and rate of turnover category. For goods retrieval, orders are received for certain product line numbers in stated quantities. Normally, goods stored the longest are dispatched first.

Locations can be recorded by means of cards moved between two boxes. In one box (the 'empty location box') cards are fitted in location number sequence while in the other box (the 'full location box'), each card is filed with the respective product line number card which follows a numerical order.

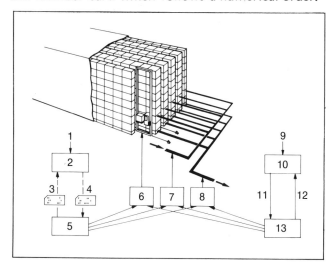

Semi- or fully-automatic remote control of stacker crane.

1 Off-line deposit and retrieval order.
2 Administrative computer.
3 Confirmation of order carried out.
4 Batch transfer (say, once every hour) of punched card information to control console.
5 Control console with card reader.
6 Controlling stacker crane.
7 Controlling input conveyor.
8 Controlling output conveyor.
9 On-line storage and retrieval order.
10 Administrative computer.
11 Continuous data transfer.
12 Continuous feedback.
13 Control computer.

Information transfer means the transfer of instructions (for example, 'Collect load at co-ordinates x12, y4, z3') from the control system to the crane. With **manual and semi-automatic operation**, (a) to (c), the crane operator does this by operating the controls or information reader by hand. With **automatic operation**, (d), there are two alternative methods of transferring signals:

↻ Continuous connection between an external control system and the crane through a signal cable, a conductor rail or an induction loop.

● Entering operating instructions when the crane is at 'home base' (accumulation method). Signals are transmitted inductively or optically (via photo-electric cells).

Being continuous, the former system is advantageous where an unspecified home base is preferred or where internal relocation of loads in the pallet racking is necessary, without the need for re-instructing the crane (for instance, moving goods forward at night). However, the accumulation method is simpler to install and operate. The crane may also be loaded up with a number of orders which, to a great extent, eliminate the need for unproductive movement back to the point at which instruction is provided.

Execution consists of moving the crane to the selected address and implementing the required transaction (order picking, pallet deposit/retrieval).

With manual control, (a), the operator is in charge and the operation may be facilitated with automated coarse or fine positioning systems. With

the former, the address is set by the operator who performs fine positioning himself. In the latter case a reverse procedure is used.

An automated crane has its own control system in the form of a microcomputer which supervises travel and lift, fine positioning, platform height adjustment and fork motion on the basis of instructions given. Travel and lift are normally two separate movements. **Coarse positioning** involves a pulse counter which checks signals between the crane mast and the racking verticals. Another system uses the microcomputer to record the movement of the driving wheel and calculate the distance covered. To be on the safe side, there is then also a calibration unit which counts rack verticals horizontally and fixed cams vertically.

When the crane approaches the required location, it automatically slows down to inching speed and is finally guided to the location by the fine positioning control.

For **fine positioning**, reflecting tape is attached to the racking beam and centred exactly below each pallet location. The crane has two pairs of photo-electric cells which scan the reflecting tape in three stages, controlling the crane through the microcomputer and steering it to the exact location. The crane has two pairs of photo-electric cells which scan the reflecting tape in three stages, controlling the crane through the microcomputer and steering it to the exact location

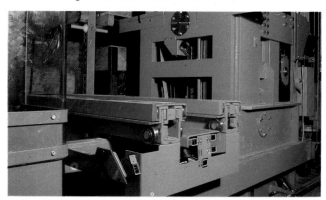

Coarse horizontal and vertical positioning is carried out using a microcomputer which counts pulses from pulse generators (such as rack uprights, reflectors, etc.). Pulse scanning sensors are situated under the forks.

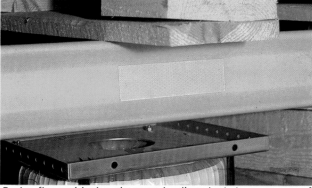

During fine positioning, the crane is adjusted relative to an area of reflecting tape on the racking beam under each pallet location. This eliminates errors caused by any beam deflection.

Such systems achieve great precision (±5 mm). The pulse generator for horizontal positioning continuously corrects its pulses to compensate for driving wheel wear, and fine positioning is always precise irrespective of deflection in the racking.

Cranes for handling pallets in racking

Even if the objective is for pallet loads to move intact from production through storage and then to the customer, order picking within warehouses is still common. It may take place within the storage space ('man to materials') or at fixed stations outside ('materials to man'). Suitable cranes for the former technique are discussed in the section 'Order picking cranes'.

Cranes for handling pallets in racking are used primarily in semi- or fully-automated warehouses where the goods are removed straight to the dispatch department or to a fixed picking station. In the latter case unused items are re-stored.

Pallet cranes normally engage pallets from the short side. The forks are approximately 60 mm thick, and the gap between the pallet decks (100 mm) is therefore sufficient for such an operation. (Previously, additional 10–50 mm spacers were required). In the United States and Japan four-way entry pallets are widely used for drive-in block stacking racks. Fork thickness is then unimportant as the forks do not enter the pallets.

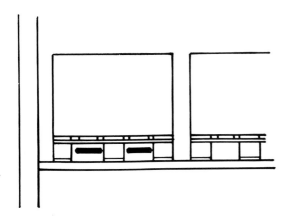

Pallets handled from the short side are supported by extendable forks which enter the gaps between bearers and base boards.

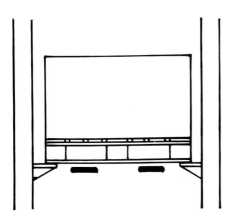

Ample space for the crane forks is provided beneath the pallet in drive-in, block stacking installations.

Stacker cranes handling pallets usually provide the highest performance figures for travel and lift. In some automated warehouses, the cranes travel at 180 m/min and lift up to 39 m.

Cranes for specialized warehouses

Stacker cranes handle not only 800 × 1200 mm Euro pallets but also:

- 800 × 2400 mm containers.
- 1200 × 2400 mm containers.
- 48 × 65 in. or 15 × 15 in. aluminium tool trays, etc.
- 1200 × 1600 mm marine pallets.
- Cassettes for lengthy loads.
- Containers for 1200 × 1700 mm metal doors held upright.
- 20 × 8 ft ISO containers.
- 400 × 200 mm cases of books.
- Specialized pallets associated with machining in automated multi-operation installations.

The same crane can accommodate loads of different sizes with one or two pairs of forks being used simultaneously. In live storage, the cranes are usually equipped with roller tables instead of forks. Some cranes carry shuttle trucks which move into the racking to park their loads.

Cranes may also deposit loads on associated roller conveyors. In this way, automated warehouses may be linked directly with production processes, thereby helping to reduce the amount of labour involved in the latter area.

This crane operating in an automated store feeds workpieces on specialized pallets to a variety of production systems.

Pallet picking and order picking cranes

Pallet picking cranes are different from **order picking cranes**. The former are equipped with forks to accommodate full pallet loads. Picking is carried out manually by the crane driver.

To avoid excessive fork reach, pallets are today handled on the long side, giving the operator a working depth not exceeding 800 mm. This environmental improvement, however, carries a penalty in that the cost of a pallet location may rise (by 50 to 100 per cent) and the aisle length for a given number of pallets will increase.

Order picking cranes have no forks but are instead equipped with plain work tables, ball tables, racks, roller conveyors or similar devices.

Pallet picking cranes operate up to 12—14 metres, and order picking cranes to 6—10 metres. Travel speeds are moderate, 90—120 m/min. Work table height can be individually adjusted to ensure a convenient working height for the driver.

Crane with load platform adjustable for height.

For comfort and safety, order picking crane cabs have anti-noise insulation, heating and 'dead man's handle'. With most makes, the driver is comfortably seated, but in some cases, he can choose whether to sit or stand while operating the crane. Another option is an inclined position, with the driver leaning against a simple bicycle seat; this is said to provide the best long term working position. Cabs are designed for rear and front entry, depending on mast location.

Some order picking cranes incorporate several picking tiers or allow several operators to work from one cab. Normally, the operator moves the picked goods to the end of the aisle, but this may be performed by paternosters or chutes feeding a belt conveyor to a sorting station. Another variant is to have belt conveyors at different racking levels.

A spiral chute for continuous removal of picked goods.

Optimizing crane movements and crane capacity

Stacker cranes represent a considerable investment and should be utilized to the maximum. To avoid unproductive travel wherever possible, input and output operations may be combined into a **dual cycle**. A rough estimate based on handling Euro pallets along aisles up to 100 m long, from racking up to 20 m high gives a dual cycle time of 3 minutes, or 20 dual cyles per crane per hour at 80 per cent utilization.

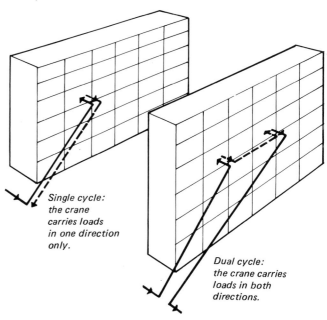

Single cycle: the crane carries loads in one direction only.

Dual cycle: the crane carries loads in both directions.

When calculating crane capacities more exactly, average travel distances over extended periods are taken into account. For instance, there are German standards (FEM 9.851) based on certain mean figures for travel, pivoting and fork travel. These standards give exact definition based on dual cycles with input at co-ordinates 1/5 L, 2/3 H and output at co-ordinates 2/3 L, 1/5 H. For capacity checking there are also alternatives defined by means of single cycles as well as input and output at different racking locations.

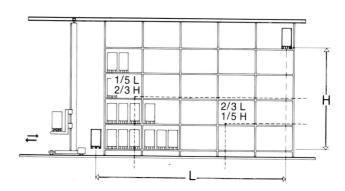

1/5 L
2/3 H

2/3 L
1/5 H

H

L

Capacity calculation based on average travel distances according to FEM standards may be used for delivery inspection.

The calculations are based on the assumption that goods are stored at random in the first empty rack opening (random location). By examining the output rate of goods from store, however, it is possible to shorten the average travel distance by storing high rate goods close to the end of the aisle.

As stated, the aim is to make the storage height/length ratio equal to the lift/travel speed ratio. As lift and travel movements are simultaneous, the crane will take a diagonal path, thereby improving access time. By locating the input point halfway up the store, two diagonals may be obtained, enabling more locations to be reached within an identical period.

Crane transfer between aisles

In some automated warehouses, storage volumes and handling rates are such as to require more aisles than stacker cranes. A transfer trolley is then used to move cranes from aisle to aisle. In large installations transfer trolleys can also be used to introduce a standby crane where a regular unit requires repair. In the earliest automated warehouses (end of 1960s), roof mounted transfer trolleys were used; nowadays floor mounted trolleys are the most common. The latter often require a pit approximately 300 mm deep and 5—6 m wide, although versions not requiring pits are now available, simplifying both installation and operations.

Transfer trolleys are usually parked at the rear of the warehouse where they also have a part-time role as guides during operations at the end of the racking. Alternatively, they can be located at the front. Tracks to accommodate the cranes can be laid directly on the trolleys. Installing guide arrangements at both aisle ends will reduce the potential storage area by a negligible amount but transfer time will be shorter since unproductive travel for aisle change is avoided. The actual transfer may be handled manually or by an automated system and takes 20—30 seconds.

Transfer trolleys, like stacker cranes, run on rails laid directly on the floor. The rails are bolted at approximately 1 metre centres and may also be recessed into concrete, especially where noise reduction is required.

Safety standards

Detailed safety instructions are to be found in the previously mentioned standards publication 'Stacker cranes' (ASS Directions No. 34) which specifies steel structures, lift and travel mechanisms, cabs, servicing routines, etc., for stacker cranes, order picking cranes and transfer trolleys. Enclosures, electrical interlocks and buffers for the cranes are also included. With regard to cab design, regulations cover control positions. Cabs must have emergency facilities by which crane drivers may leave the cab through special panels in the event of breakdown. Masts must also be equipped with steps.

Safeguards against attempts to deposit loads in occupied rack space must be incorporated. Such equipment may consist of a microswitch mounted on the forks — like the tongue of a snake — which senses whether or not a given location is empty. Another common method is to fit a friction clutch to the extendable forks so that, if anything obstructs the load, the clutch will slip and sound an alarm. Nowadays infrared sensitive photoelectric cells check the availability status of locations.

To avoid skewing loads, cranes are fitted with photoelectric cells which detect overhangs and activate an alarm.

Finally, safety requirements also state that only competent operators must drive cranes. They should be provided with safety keys, preventing unauthorized crane operation by other personnel.

Precise co-ordination with other systems

A satisfactory automated warehouse demands precisely manufactured components and equally precise assembly (rack uprights, for instance, must not be out of plumb). Manufacturers provide guidance on the tolerance limits required for accommodating assumed lateral and vertical deflections (75—100 mm).

The figures below indicate approximate tolerance limits:

- Position of uprights ±5 mm.
- Deflection of uprights (at 15 m height, approx.) ±25 mm.
- Max. beam deflection 1:300 (e.g., a 2 m beam may have a 6.7 mm deflection).

The above limits pre-suppose sensing against a point of reference (reflecting tape), otherwise closer limits are required.

To achieve co-ordination between transport systems and cranes, loads must normally be aligned against fixed stops or bars prior to their arrival at the collection point.

Transport systems for goods input and output

Stacker cranes handle loads via separate input and output lines (pick—and—place stations). Both lines are usually situated at the front end of the storage, but in some cases goods output is at the rear. Pick—and—place stations have recesses for crane forks and are often equipped to align loads. Several different transport systems can be used:

- Roller conveyor/chain conveyor.
- Roller conveyor/rail mounted trolleys.
- Rail mounted trolleys/roller conveyor.
- Rail mounted trolleys exclusively.
- Fork trucks.
- Detachable trolleys towed by an underfloor chain conveyor followed by manual transfer to the picking position.
- Vehicles controlled by induction loop.
- Overhead conveyors.

The input and output systems may be located at the same level or one above the other. Where systems share the same level, some elements may be common between input and output, reducing total costs but increasing utilization (waiting may be necessary).

Crane performance will increase slightly as only fork movement is needed to switch between outgoing and incoming goods. Where the pick—and—place stations are situated one above the other, a crane lift movement is added. However, this layout helps increase storage space, and an additional pallet rack, for instance, can replace the pick—and—place stations.

Loads are guided to the selected crane by a special control system or by the high level storage mini-computer. Data zones help ensure the loads arrive at the correct address (see special section on control under main heading 'General information on line restricted materials handling equipment').

Roller conveyors and chain conveyors usually operate at 8—10 m/min. For accumulation purposes, the lines may be sub-divided or accumulation features may be incorporated into the roller conveyor.

The input/output system in a high level storage is a fairly costly item and therefore requires exhaustive analysis. Computer simulation can often be employed to check the capacity of different systems.

Pick—and—place stations may be situated at either side of an aisle or one above the other.

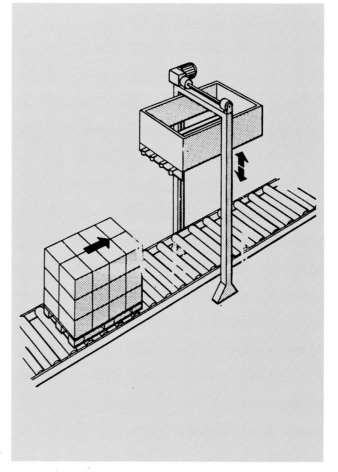

The load profile and symmetry of inout pallet loads should be checked carefully to avoid subsequent malfunction. Rejected loads should be removed and built up again.

STACKER CRANE VOCABULARY

The stacker crane vocabulary includes the following terms, rendered in Swedish, English and German:

Höglager	High bay warehouse	Hochregalanlage
Lastpall	Pallet	Palette
Modulsystem	Modular system	Bausteinsystem
Transportgång	Aisle	Gang (Gasse)
Hämta-lämna-läge	Pick-and-delivery station (P&D)	Ein- und Ausschleus-station
Insättning-uttagning	Store-retrieve	Einlagern-Auslagern
Staplingskran	Stacker crane	Regalförderzeug (RFZ)
Plockkran	Picking crane	Kommissioniergerät
Överföringsvagn	Transfer car	Umsetzer (Umsetzbrücke)
Teleskopgafflar	Telescopic forks	Teleskopgabeln
Dubbelcykel	Dual cycle (dual command)	Doppelspiel
Cykeltid	Cycle time	Spielzeit
Åkhastighet	Travel speed	Fahrgeschwindigkeit
Lyfthastighet	Hoist speed	Hubgeschwindigkeit
Gränslägesbrytare	Limit switch	Endschalter
Datorstyrning	Computer control	Rechnersteuerung
Processdator	Process computer	Prozessrechner
Hålkortsstyrning	Punch card control	Lochkartensteuerung
Hängkabel	Festoon cabel	Schleppkabel
Strömskena	Bus bar	Schleifleitung

Installation examples

These pictures show examples of stacker crane installations
(see also 'Case studies').

Storage for lengthy loads.

Picking from shelving.

Pallet storage.

EXAMPLES OF STACKER CRANE AND ORDER PICKER SPECIFICATIONS

	Order picker crane	Stacker crane H = 15 m	Stacker crane H = 25 m
Travel speed	90–120 m/min.	90–180 m/min	90–180 m/min.
Elevation speed	10–30 m/min.	10–40 m/min.	10–40 m/min.
Acceleration	0.3 m/sek^2	0.3–0.5 m/sek^2	0.3–0.5 m/sek^2
Aisle width requirements (stacker crane, pallets handled on short side)	1,000–1,200 mm	1,500 mm	1,500 mm
Deadweight	3,000 kg	6,000 kg	9,000 kg
Power input	8 kW	20–25 kW	20–25 kW

General information on line restricted materials handling equipment

Line restricted materials handling equipment moves goods continuously or intermittently along a predetermined route. Such equipment can usually be considered as a fixed installation — an essential difference compared with unrestricted equipment such as forklift trucks — and cannot be as easily adapted to changing conditions for internal goods flow. Also, the line frequently represents a physical obstacle to the movements of vehicles and personnel.

However, the extent of this obstacle varies greatly, depending on whether the line is floor or roof mounted. Some lines (such as induction loops buried in the floor) present no obstacle at all.

Line restricted equipment is suited mainly to activities involving fairly large, predictable goods flow. The degree of flexibility varies for different types of equipment and may be influenced in several ways, e.g., by alternative addressing and accumulation features.

When addressing, sorting, goods assembly and similar activities are mechanized, some type of line restricted equipment is usually inevitable. The types most commonly used (e.g., roller and belt conveyors) are available as kits based on a number of standard components. They provide ample opportunities for individual layouts, avoiding the need for special designs.

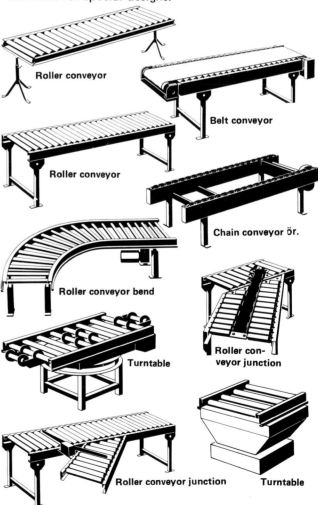

Roller conveyor

Belt conveyor

Roller conveyor

Chain conveyor ör.

Roller conveyor bend

Turntable

Roller conveyor junction

Roller conveyor junction

Turntable

Examples of standard components for line restricted transport systems.

Line restricted transport system.

Classification of line restricted mechanical handling devices

Line restricted handling devices may be classified according to how goods are moved.

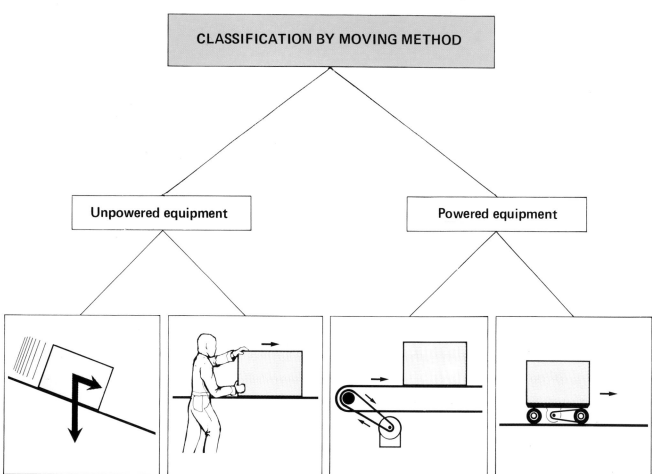

CLASSIFICATION BY MOVING METHOD

Unpowered equipment

Powered equipment

Gravity operation *means that the line is inclined and friction (rolling or sliding) is low enough for the weight of the goods to effect the movement required. One problem is to achieve a controlled, well adjusted speed which can accommodate goods of different weights. (More about this in the 'Roller conveyors' section.)*

Manual operation. *Unpowered lines may also be mounted horizontally. Goods are moved by hand or by some other external means.*

Line powered operation. *A stationary source provides the power. Goods rest on or are suspended from the line. The line may also be divided into several sections, each individually powered.*

In a **unit powered** *operation the line serves merely as a guide path for mobile units or load carriers equipped with their own propulsion units. Such powered units usually have the advantage of being detachable, providing greater flexibility than line powered operations. In addition, designs are generally simpler and easier to modify. Mobile units or load carriers are, however, rather more complex and expensive.*

A more usual method of classifying line restricted mechanical devices is to refer to the design of the load carrying components.

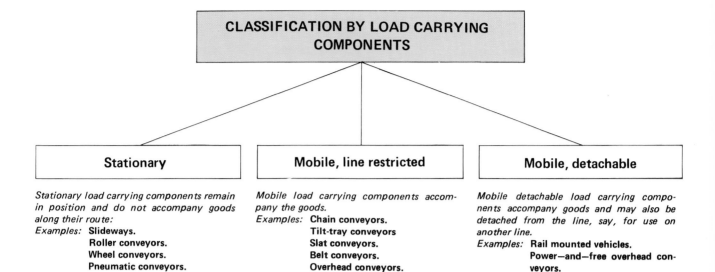

CLASSIFICATION BY LOAD CARRYING COMPONENTS

| Stationary | Mobile, line restricted | Mobile, detachable |

Stationary load carrying components remain in position and do not accompany goods along their route:
Examples: **Slideways.**
Roller conveyors.
Wheel conveyors.
Pneumatic conveyors.

Mobile load carrying components accompany the goods.
Examples: **Chain conveyors.**
Tilt-tray conveyors
Slat conveyors.
Belt conveyors.
Overhead conveyors.

Mobile detachable load carrying components accompany goods and may also be detached from the line, say, for use on another line.
Examples: **Rail mounted vehicles.**
Power—and—free overhead conveyors.
Automatically guided vehicles.

If classification by moving method is represented by one co-ordinate and classification by load carrying components by the other, the result is a diagram like the one below. The diagram may be used to indicate those different combinations which are used in practice.

| Classification by moving method | Classification by load carrying components | | | | | | | | | | | |
| | Stationary | | | Mobile, line restricted | | | | | Mobile, detachable | | | |
	Chute	Roller conveyor	Wheel conveyor	Belt conveyor	Overhead conveyor	Chain conveyor	Tray conveyor	Slat conveyor	Rail mounted vehicle	Self-powered overhead conveyor	Overhead conveyor power—and—free	AGV
Unpowered, gravity operation	●	●	●		●				●			
Unpowered manual operation		●	●		●				●			
Line powered operation		●	●	●	●	●	●	●	●		●	
Unit powered operation									●	●		●

CLASSIFICATION BY CONTROL METHOD

Line restricted handling devices may also be classified by control method. There is a distinction between mechanical and electromagnetic control. The latter usually employs a conducting loop buried in the floor. Changing the layout of such systems is usually easier than changing a mechanical control system.

Characteristics and assessment criteria

Assessing the suitability of line restricted handling equipment for an application demands extensive knowledge of the equipment's characteristics.

A useful method is to draw up a checklist of the major characteristics. The table below provides an example of typical parameters. There are three main groups: operation, technical quality, ergonomics and environment. (With regard to costs, see section 'Calculation methods and forms of financing'.)

Checklist for line restricted handling devices	
Operation	**Ergonomics and environment**
Installation method: In the floor. On the floor. Overhead.	**Adaptability:** Possibility of varying line direction and length. Possibility of increasing capacity. Effect of other adjacent work functions.
Direction of movement: Vertical. Horizontal. Inclined.	**Maintenance.**
Relevant features of goods carried: Weight. Shape. Size of base area. Nature of base area.	**Reliability.** **Energy consumption.**
Performance: Capacity. Speed. Acceleration.	**Ergonomics and environment**
Possibilities of accumulation	Noise. Exhaust emission. Vibration. Working position. Physiological strain. Accident risks. Other risks. Unimpeded views.
Possibilities of loading: Continuous. Intermittent.	
Route switching: For movement variations. For sorting.	

Some items above are especially important for line restricted handling devices and may be worth a more detailed explanation:

Installation method is partially determined by the demands of other transport systems. Where the need is for unimpeded work, with powered and unpowered trucks complementing line restricted equipment, an overhead installation may be the best choice. The decision also depends to a great extent on the inclines involved.

Overhead conveyors carrying suspended loads are generally more suited to sequences in which loads have to be both lowered and raised. On the other hand, where goods need only be moved from higher to lower levels, floor and frame mounted chutes can be used if the height differences are moderate.

Accumulation is essential if delays occur in different places in a transport system or if accumulation itself is a necessary input/output element (e.g., a buffer stock for distribution vehicles). Accumulation is simplest with unpowered wheel or roller conveyors, the only necessity being some method of stopping the goods flow. With powered conveyors, accumulation is possible in two basically different ways. Goods can be stopped while friction between goods and conveyor or between rollers/belt and drive is allowed to continue. Alternatively, part of the line may be disconnected from the drive. Further details of accumulation techniques are given in the subsequent section which describes the different types of equipment.

In some cases line restricted handling equipment, may be **loaded** at any time (e.g., with roller and belt conveyors). In other cases a load carrier may have to arrive first (e.g., a platform or truck).

Route switching, i.e., directing goods onto other lines, is important where sorting is required (e.g., at transhipment terminals, central depots). Switching may be by hand or mechanical. In the latter case, automatic addressing and sorting are possible. (See section 'Controlling line restricted handling devices').

Adaptability. The role of a given item of handling equipment may, of course, change considerably before the equipment has reached the end of its lifespan. Capacity requirements may have increased, different goods may now be carried. Warehouse layout and production installations may be changed. Adaptability is the quality of adjusting to such changes.

Controlling line restricted handling devices

A simple conveyor line for moving goods between two points needs no special control equipment: it is controlled by switching the power unit on or off.

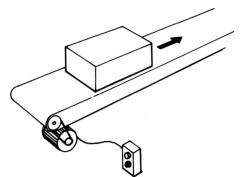

The simplest form of control: an on/off switch.

But where several types of line have been combined (roller and chain conveyors, overhead conveyors, rail mounted vehicles, etc.) an overriding **control system** is required, to direct goods as desired. This means that route data are needed at each switching point.

Such data (control signals) may be provided manually or mechanically according to the following alternatives.

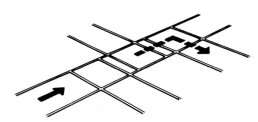

The control system changes the direction of both supported . . .

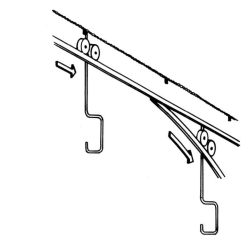

. . . and suspended goods.

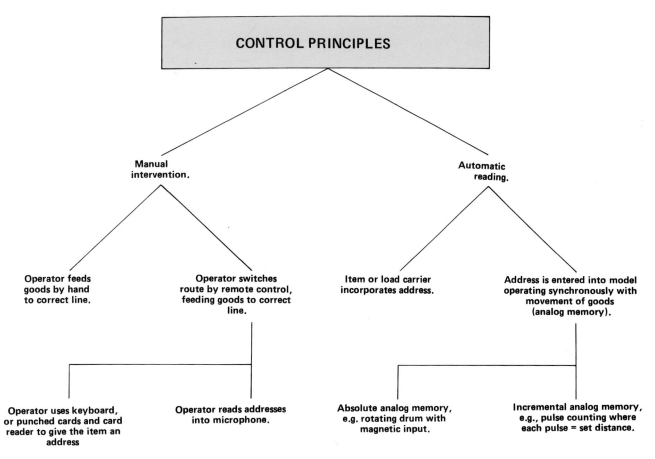

CONTROL PRINCIPLES

Manual intervention.

- Operator feeds goods by hand to correct line.
- Operator switches route by remote control, feeding goods to correct line.

Operator uses keyboard, or punched cards and card reader to give the item an address

Operator reads addresses into microphone.

Automatic reading.

- Item or load carrier incorporates address.
- Address is entered into model operating synchronously with movement of goods (analog memory).

Absolute analog memory, e.g. rotating drum with magnetic input.

Incremental analog memory, e.g., pulse counting where each pulse = set distance.

An **operator controlled system** with keyboard/punched cards sends pulses to electromagnets which may open a hydraulic cylinder valve or initiate an electric motor drive. In both cases, linear motion lifts a table, positions a guide, switches a rail section, etc.

In **automatic reading**, the address is carried in some form on the item label or on the load carrier, e.g.:

- Label with black and white squares. The squares normally correspond to a binary coded number and are read by photoelectric cells, one cell for each square. The possible combinations of addresses depend on the number of squares: n number of cells provide $2^n - 1$ possible addresses

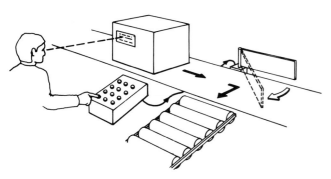

Operator controlled, keyboard operated system.

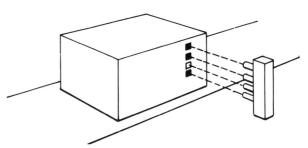

Binary coded address consisting of black and white squares.

(e.g. if n = 4, the number of addresses is $2^4 - 1$ or 15).

- Barcode label standardized as EAN and UPC codes and often used on many everyday articles. It is read with a small light pen.

With a **voice addressed system** each operator pre-records simple words (one, two, three . . . stop, start, etc.) on a magnetic tape. At the start of the day, the operator places 'his' tape in a tape recorder connected to the control system which is able to 'translate' spoken instructions into pulses and thus generate the required control signals. Voice addressed systems are useful where goods of various size arrive simultaneously on a line, making it necessary for the operator to use both hands to lift and turn each item to read the address (e.g., postal parcels).

EAN barcode for identifying merchandise may also be used for addressing purposes on load carriers.

- Flags read by mechanical switches (microswitches), photoelectric cells or magnets. As with the system mentioned above, $2^n - 1$ addresses are possible with n number of flags on the load carrier. When choosing between control systems, it should be remembered that microswitches are cheaper but non-contact sensors are more reliable since there is no mechanical wear. However, dead lightbulbs or dirty reflectors in photoelectric systems can be a problem.

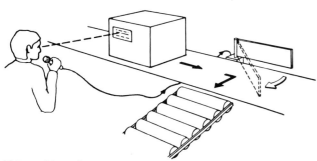

Voice addressed control. Usually a microphone is used, leaving the hands free to orient parcels for reading.

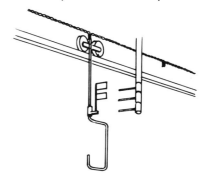

Routing system with mechanical switches is cheap but requires more maintenance because of wear.

- Labelling the conveyor. With steel belt conveyors, addresses may be imposed electromagnetically on the belt itself on a spot immediately in front of each goods item. The information is picked up by a reading head and then erased during the belt's return (cf. section 'Belt conveyors').

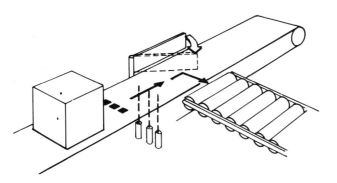

Routing information imposed electromagnetically on a steel belt conveyor.

Automatic systems with **analog memory** previously functioned on the absolute principle, for instance, via a rotating magnetic drum, or a rotating disc and movable cams (or balls). When the drum had rotated through an angle which corresponded to the goods having moved to the switching position, an electromagnetic signal on the drum actuated a transducer which generated a switching signal.

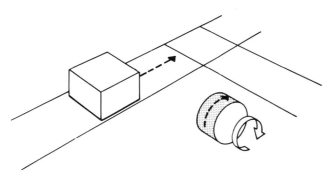

Analog memory with synchronously rotating drum.

Nowadays **incremental** analog memories are normally used, based on shift registers or microcomputers.

These systems function in the following manner. First the line is divided into modules (data zones), each zone having a transducer (microswitch, photoelectric cell). One goods item only may be present in each zone. Items start in zone 0 where their identities (and thus their addresses) are established and fed into the memory by an operator. As goods move through each data zone, the appropriate transducer sends an acknowledgement

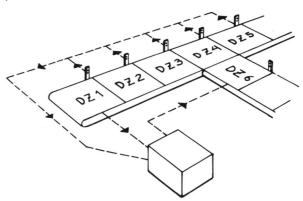

When an analog memory is combined with a computer, the line is normally divided into zones. A microswitch or photoelectric cell acknowledges when a goods item has entered a new zone.

signal enabling the computer to move the goods forward by increments in its memory.

As mentioned above, both microswitches and photolectric cells can be used to send signals. The contacting member of a microswitch can easily be damaged, say, by a pallet. As a result, indirect actuation is often preferred, in the form of a metal flag which, when read, provides a signal to initiate the microswitch. This increases the cost, however, and makes photoelectric cells a more attractive choice. (Note: flags may also move, producing two signals instead of one.)

In some awkward zones (for instance, in systems which must be capable of accommodating half-pallets as well as pallets) photoelectric cells must nevertheless be used for reasons of space.

Systems may evolve gradually, but where there are many functions switching sequences may advantageously be studied and formulated using Boolean algebra. In this way each function can be defined by an algorithm which can then be simplified. This reduces the number of signalling devices.

Roller conveyors

Roller conveyors are line restricted devices and consist of rollers mounted between two side members. Bearings are usually incorporated in the rollers to cut down mechanical losses.

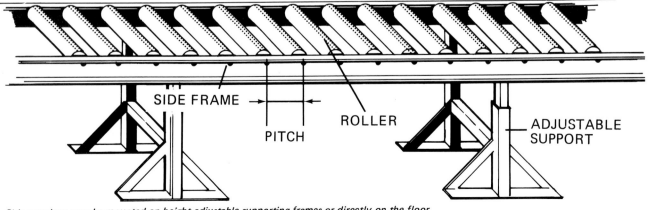

SIDE FRAME →

PITCH

ROLLER

ADJUSTABLE SUPPORT

Side members may be mounted on height adjustable supporting frames or directly on the floor.

Roller conveyors may be **unpowered** or **powered**.

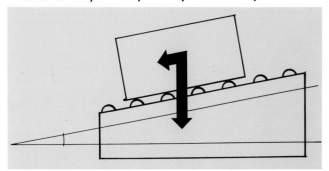

An **unpowered gravity roller conveyor** is set at an appropriate incline, and goods move down it by gravity.

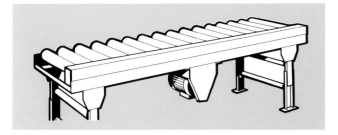

In **powered roller conveyors**, the power unit, normally an electric motor, drives the rollers via chains or belts, providing controlled horizontal movement of goods.

Application

Roller conveyors are among the most commonly used forms of line restricted internal transport. They can be used for many different types of goods, from small cardboard boxes up to pallet loads and even larger items. Goods must have flat bases or be fitted with skids in the direction of the movement. Drums and tins with seamed bottoms can also be conveyed if the conveyor is equipped with guide rims. However, roller conveyors are less suitable for moving easily deformed goods.

As a rule, roller conveyors function best if the underside of the goods they transport is firm.

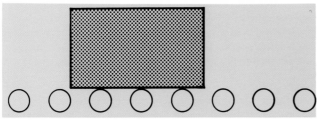

At least three rollers should support each load.

Distances between rollers must be adapted to the smallest items carried. As a rule, items should be in contact with at least three rollers simultaneously.

Unpowered roller conveyors are used over distances short enough not to create problems at inclines and where there are no speed control requirements.

As they allow accumulation and are, at the same time, relatively simple and cheap, unpowered roller conveyors are extensively used in industrial warehouses and production installations and in commercial depots and stores.

The most common applications are in goods reception and dispatch where distances are normally short and accumulation requirements critical.

When pallet loads or boxed goods are received at an industrial warehouse, they may be unloaded from the delivery vehicles by truck or by hand and transferred to an unpowered roller conveyor which moves them to an accumulation point prior to unpacking and checking.

Corresponding operations apply to loading vehicles with goods taken out of storage. There are similar applications in various industrial processes where transport and accumulation requirements coincide, for instance, when taking and moving materials from intermediate storage to a production line or vice versa (see section 'Application examples').

Live storage is currently attracting interest, i.e., block stacking racks for pallets or boxed goods, equipped with unpowered roller conveyors (or wheel conveyors, see below).

Applications for unpowered roller conveyors are limited by the incline needed to move goods. Differences in levels between input and output may cause practical difficulties and limit transport distances.

Also, handling speeds generally vary with the weight and nature of the goods. Items of constant weight and with similar undersides give best results. The table indicates appropriate inclines for some types of goods unit.

Goods type	Appropriate incline
Heavy pallets and wooden bins	2.5–3.5%
Cardboard boxes 10–20 kg	3–5%
Cardboard boxes, 0.5–1 kg	10–13%

These figures indicate difficulties that may occur where weights vary. Some items may reach too high a speed, others will remain stationary. The same may occur where undersides have different characteristics or pallets are damp. If the variations are great, it may be difficult to arrive at an incline that meets a specific goods flow. To control speeds more satisfactorily, braking rollers have been developed to apply deceleration relative to the speed of the rollers (see also section 'Design').

Powered roller conveyors are used for the same types of goods and mainly for the same transport tasks as unpowered conveyors. A drive becomes essential where one or more of the following requirements must be met:

- Long transport distances.
- Horizontal transport.
- Controlled speed.
- A high degree of automation with respect to route switching, sorting, etc.

Typical applications are:

- Moving goods for storage at high levels.

- Transporting picked units from storage to packing and dispatch, or from storage to sorting.
- Transporting fixtures or objects between assembly stations in a production line.
- Moving goods along a packing line (pallet charging, shrinkwrapping, strapping, dispatch).

Unlike unpowered roller conveyors, powered types will not automatically provide an acceptable accumulating function and various special solutions are needed. There are two basic methods. One is to allow one or several rollers to slip (beneath the goods or relative to the drive). The other is to divide the line into sections, each capable of being disconnected from the drive when required (more about this in section 'Accumulation').

Design

Rollers usually have steel casings and shafts supported by ballbearings (see figure).

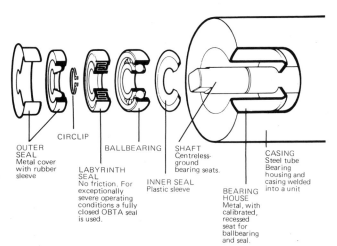

OUTER SEAL
Metal cover with rubber sleeve

CIRCLIP

LABYRINTH SEAL
No friction. For exceptionally severe operating conditions a fully closed OBTA seal is used.

BALLBEARING

INNER SEAL
Plastic sleeve

SHAFT
Centreless-ground bearing seats.

BEARING HOUSE
Metal, with calibrated, recessed seat for ballbearing and seal.

CASING
Steel tube
Bearing housing and casing welded into a unit

Casings may be coated externally with plastic or rubber for better friction and lower noise levels. Casings entirely made of plastic are sometimes used for moving lightweight items where hygiene is at a premium, e.g., in the food industry.

Cardboard cased rollers are for moving lightweight goods.

Types of rollers.

Rollers are usually held between steel or aluminium side members which are mounted on height adjustable legs or directly on the floor.

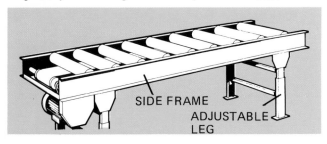

SIDE FRAME

ADJUSTABLE LEG

Unpowered roller conveyors may be fitted with braking rollers for better speed control, e.g., when goods items vary in weight.

The basic principle is that braking action should increase with speed.

Customary solutions are:

- Eddy current brake — aluminium disc rotating in a magnetic field.
- Hydraulic brake — internal friction via hydraulic fluid.
- Centrifugal brake — shoes press outwards against a stationary drum when a preset number of revolutions is exceeded.

If braking rollers are used, an incline may be selected to provide sufficient speed for the lightest items while heavier items are restrained by the brake.

Brake rollers can be mounted under the conveyor line to provide indirect braking or may replace conventional rollers at suitable intervals.

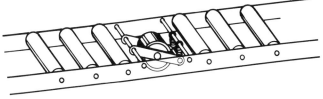

Directly acting braking roller.

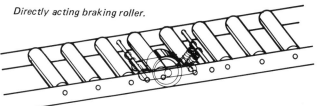

Indirectly acting braking roller.

An alternative method of achieving slow, controlled movement is to impart a reciprocal motion to the entire conveyor line. This prevents goods from remaining on the conveyor even if the incline is slight. It is a relatively expensive and complex solution, however, as the entire line must be mounted on rollers and power operated. A powered roller conveyor with friction rollers may be an alternative.

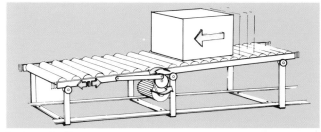

Reciprocal roller conveyor.

Powered roller conveyors are similar in basic design to unpowered versions. The drive is often by means of an electric motor through chains or belts. Belt operation is simplest and cheapest and is suitable mainly for lightweight goods. The design may involve separate drive belts for each roller or a single belt driving a number of rollers by friction from below.

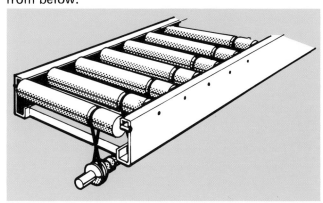

Each roller is driven by separate belt.

Rollers driven by a single common belt below the conveyor line.

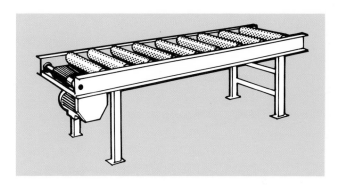

Flat belt drives the rollers from underneath.

Belt operation has the advantage of allowing accumulation on the line by a fairly simple adjustment between belt and rollers (see also section 'Accumulation').

Chain drive employs single or double chains. With the former each roller is driven via sprockets at each end.

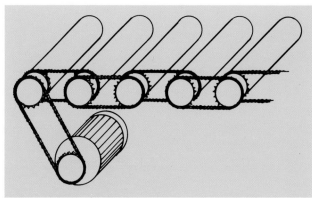

Double chain operation.

With double chain operation, each roller has two sprockets and transmission is by individual chains from roller to roller, resulting in excellent driving capacity.

System design

When devising internal transport systems which employ unpowered and powered roller conveyors, an extensive range of special equipment for stopping, diversion, switching and accumulation is available.

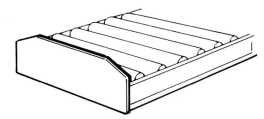

The simplest stopping device is a fixed stop . . .

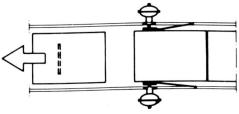

With cardboard and other boxes, two air-operated clamps will restrain subsequent items while the leading item is allowed to move on.

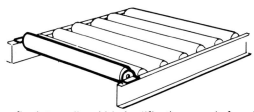

. . . or a fixed stop roller which simplifies the removal of goods from the end of the line.

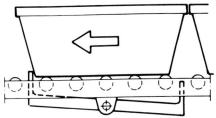

To separate boxes and bins an indexing stop can be used. Sufficient space is needed between goods items to permit the rear stop to emerge above the rollers.

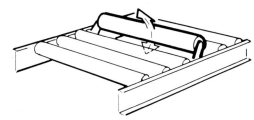

Pivoting stops are designed as either rollers or panels.

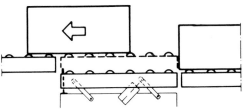

A lifting stop separates close wooden and cardboard boxes, etc., although goods items must be of roughly the same length.

With an inclined roller conveyor, a braking ramp is the simplest method of providing gentle retardation and cushioned impact when goods accumulate on the line.

Bends with tapered rollers.

Bends are available with standard angles 30°, 45°, 60° and 90°. Radii vary from 800 mm up to 2,500 mm. Rollers may be straight or tapered.

Bends with two-part rollers.

Split rollers prevent goods from swivelling in bends by allowing the outer rollers to rotate at a greater rate than those inside. This also reduces speed losses in bends.

Bend with three-part rollers for Euro pallets.

Special conveyor bends with three-part rollers are used for moving Euro pallets. The three rows of rollers coincide with the pallet's three baseboards. All types of bend may be powered or unpowered.

Various types of guide flanges guide goods through bends and straight sections.

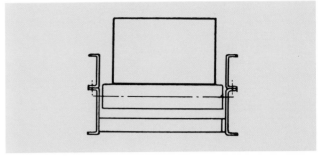

Simple guide.

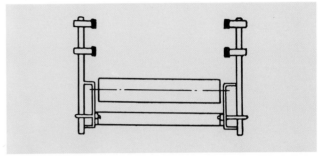

Alignment guide.

Flanged rollers for guiding pallets.

Centrally positioned flanges guiding the central baseboard are often used for Euro pallets.

Switching sections are used where several lines of goods merge into a single flow or where a flow is split into several lines, e.g., for sorting.

Ball table.

To re-route goods by hand a ball table can be used. Balls can also be mounted on height adjustable cross-members between the conveyor rollers.

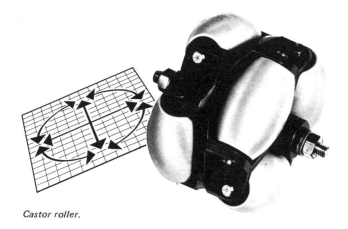

Castor roller.

Castor rollers used instead of balls, can be powered.

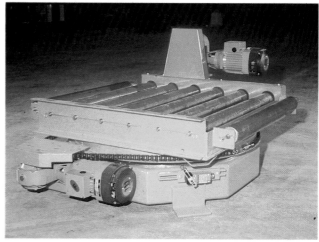

Roller conveyor turntable.

The extensive range of alternatives for right-angle switching and goods input includes a roller surface turntable capable of being rotated through 90°. Normally electrically powered, it is slower than other alternatives. On the other hand, it may be designed to handle goods with great care.

Junction using short rollers.

Switching can also be accomplished with various combinations of slewing and tapered conveyor sections.

Right-angle switching using height adjustable wheel conveyors.

Manual transfer trolley.

Transfer trolleys are very common. They move goods and pallets between parallel conveyor lines for marshalling, sorting, input and output in single tier live storage installations.

The trolleys normally run on rails and may be powered or unpowered.

Switching by roller conveyor.

Another common method is to mount chains, rollers or wheels between the rollers in a roller conveyor and to make either the conveyor or the intermediate components height adjustable for switching purposes.

Roller and wheel switching devices may be powered or unpowered. Unpowered switching is possible by raising the retractable wheel or roller sections at one end: goods then roll off by their own weight. Powered roller or wheel switching units, however, are normally designed for parallel lifting. This

Switching by chain conveyor.

method is best for cardboard or other boxes with flat bottoms.

Accumulation

For right-angle switching of Euro pallets, chain switches are commonly used. A section of a roller conveyor is lowered and the pallet is transferred onto the chain.

If a continuous goods flow is disturbed, accumulation is necessary. It may be a matter of changing from one type of handling to another (e.g., during loading/unloading or storage/retrieval). It may also be that breakdowns create a need for flexible accumulation capacities.

Unpowered, inclined roller conveyors permit accumulation, as it occurs, on the assumption that goods may be allowed to collide with and press against each other. Speeds must be adjusted to prevent damage; this may be difficult, particularly if items vary in weight or if base surfaces are different. To minimize these effects, braking rollers may retard the speeds of heavy items, or a reciprocating roller conveyor may be used.

To facilitate accumulation on powered roller conveyors, an unpowered section can be inserted,

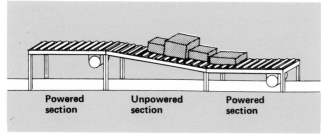

Accumulation on section of unpowered roller conveyor.

but this will of course create the above mentioned problems.

Alternatively, the entire powered line may incorporate rollers with built-in friction clutches which slip if goods are held up. Such units have well-specified friction characteristics so that compressive forces between goods items are small unless weights show very great variations.

Belt operated roller conveyors may be designed so that the belt snubs the rollers on the underside, performing a similar function. This solution is not suited to heavy goods which require chain drives.

Yet another method resulting in very low compressive forces is to mount rollers on a chain conveyor, creating a roller conveyor which accompanies goods. When goods move forward, the rollers do not operate; when goods begin to accumulate, the conveyor continues to move forward while the rollers start to rotate.

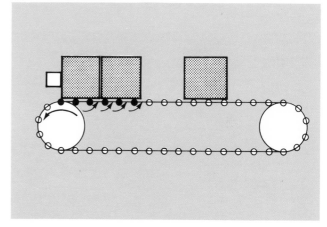

Chain operated roller conveyor. The rollers rotate under stationary goods, keeping compressive forces low.

All the accumulation methods described above are associated with some degree of compressive force and some form of friction. Where goods are particularly vulnerable to impact or vibration, this may be unacceptable and an accumulation arrangement entirely free from compressive forces may be necessary. This is achieved by dividing the conveyor line into sections. As a goods item progresses along a section of conveyor, it actuates the drive for the item behind. Actuation may be mechanical, pneumatic, electrical or photoelectrical.

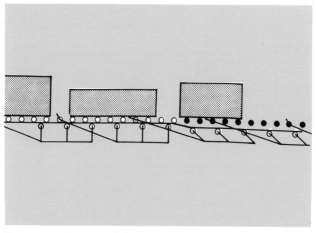

Non-compressive accumulation with sectional triggering of drive.

Application examples

The photographs below illustrate typical examples of roller conveyor application.

Internal transport of wooden bins in a central depot for specialized goods.

Bay installation for automatic loading/unloading of pallets transported by lorry.

Live storage of boxed goods.

Live storage of pallet loads.

Roller conveyor installation for moving loads.

Development trends

Roller conveyors have long been the traditional standard solution for internal transport systems with steady flows of weighty goods such as bins, wooden boxes and pallets.

There are many indications that roller conveyors will become less important as more and more competitive equipment enters the market. For instance, various types of vehicle systems may allow individually coded routing and, at the same time, simpler and less costly design/construction.

Automatically guided vehicles (AGVs) may also successively reduce the market share of roller conveyors, primarily in goods reception and dispatch and in connection with internal transport between stores, production and finished goods warehouse. AGVs offer greater flexibility with respect to individual routing and layout modifications.

Procurement checklist

Anyone considering an investment in roller conveyors has a great many factors to take into account. A practical aid is to draw up a checklist to ensure no essential aspect has been overlooked.

A checklist for roller conveyors might be as follows:

• How will the proposed layout affect other transport operations on the premises, now and in the future?

• Are split roller features and load capacities capable of being adapted to goods flows which may be modified within the conveyor's economic life?

• Has the potential of unpowered conveyors, wherever practicable, been considered?

• Have maintenance costs been taken into consideration with regard to choice of power system?

• Have the possibilities of achieving simpler and cheaper accumulation by different load carriers or different packaging been considered?

EXAMPLES OF ROLLER CONVEYOR SPECIFICATIONS

Conveyor type	Roller type	Suitable goods items	Capacity kg/m	Conveyor width mm	Standard section length m	Speed m/min.	Remarks
Unpowered, light duty	Plastic Steel	Cardboard boxes, plastic bins	50−200	200−800	3	dep. on incline	
Unpowered, medium duty	Steel	Cardboard boxes, bins, wooden boxes	200−700	200−1,000	3	,,	
Unpowered, heavy duty	Steel	Euro pallets, heavy boxes	2,000	400−2,400	Non-standard	,,	
Powered, O-belt, light duty	Plastic Steel	Wooden and cardboard boxes	5−100	200−1,000	3	8−30	Modifiable for accumulation
Powered, flat belt, light duty	Plastic Steel	Bins, cardboard boxes	5−100	200−600	Max. 12	20	Flat belt on slider bed
Powered, belt, medium duty	Plastic or steel	Wooden and cardboard boxes	100−400	300−900	9−12	10−35	
Powered, chain, medium duty	Steel	Wooden and cardboard boxes	200−400	300−600	5−20	8−20	
Chain operated heavy duty	Steel	Pallets	100−1,500	200−1,400	5−20	8−20	Non-accumulating
Chain operated, heavy duty, accumulating	Steel	Pallets, heavy wooden boxes	1,500	200−1,400	5−20	8−20	

Wheel conveyors

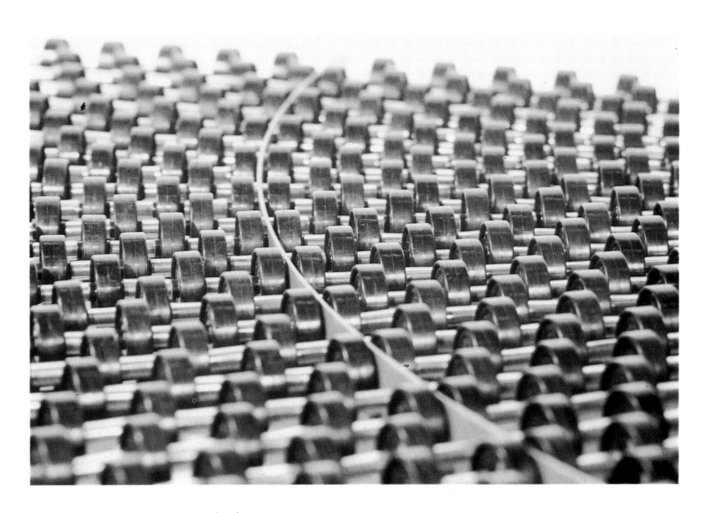

The most common wheel conveyors are basically similar to unpowered roller conveyors but incorporate wheels mounted on shafts rather than rollers. Wheels may be steel or plastic, with ball-, roller or plain bearings.

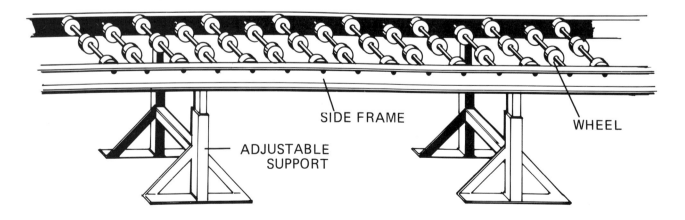

SIDE FRAME

WHEEL

ADJUSTABLE SUPPORT

Application and design

As they can be set in a staggered pattern, wheels form a closer configuration than rollers, reducing the need for flat undersides on the goods transported. Wheel conveyors can thus be used for smaller and more irregular items than roller conveyors.

This type of wheel conveyor is suitable mainly for lightweight items (less than 50 kg) such as cardboard boxes and bins of various kinds with fairly even bottoms.

However, wheel conveyors cannot stand point loads as well as roller conveyors because rollers distribute loads more satisfactorily.

Wheel conveyors are mostly unpowered and function as gravity lines, with the item's weight determining travel speed.

Applications basically coincide with those described for unpowered roller conveyors, except that wheel conveyors are applicable for lightweight goods only.

Choosing a wheel conveyor rather than a roller conveyor will depend on the size and nature of the base area of the goods. Small, lightweight items do not demand steep inclines to keep moving. (In comparison with roller conveyors, the moment of inertia of a wheel's rotating parts is considerably less.) If the route has bends, a wheel conveyor is also simpler and cheaper (roller conveyors demand tapered or split rollers). Sometimes straight roller conveyors are combined with curved wheel conveyor sections.

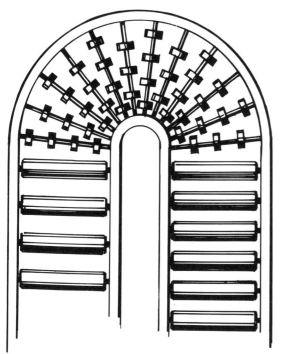

Combination of straight roller sections and curved wheel sections.

Wheels may also be mounted in profiled sheet metal.

Wheel conveyor with plastic wheels mounted in profiled sheet steel.

The result is a comparatively simple and low cost conveyor for lightweight goods (under 20 kg). Wheels are then usually made of plastic.

The figure below shows another simple type of wheel conveyor designed, for example, for industrial do-it-yourself assembly.

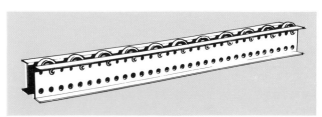

Wheel conveyor with plastic wheels mounted on steel tubing bent to form the required curve.

This is constructed of plastic wheels with simple steel mountings on steel tubing. The tubing can be bent as required.

Heavy goods and pallet loads require the wheels to be mounted in side frames.

Wheels may be installed in a single line . . .

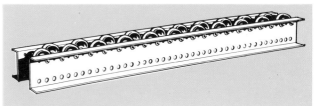

. . . or in twin staggered lines.

143

Two or more side frames may be mounted in parallel at suitable distances.

Wheel conveyor with two parallel side frames.

Thanks to the close positioning of the wheels, pallets may be moved with their baseboards placed transversely. Such conveyors may complement a transport system where roller conveyors are used to transport pallets with baseboards placed in the direction of movement. These wheel conveyors are available in standard designs for loads of 200—1,000 kg per wheel.

With heavy pallet loads and pallets of uneven quality, however, it may be difficult to achieve controlled even speeds down inclined conveyors. To counter this difficulty, a special type of wheel (Palletflo) has been developed. The wheels deform to provide a built-in braking action.

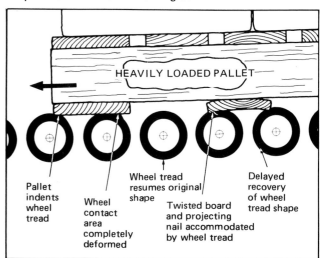

This type of conveyor can be used with advantage in live storage installations and in long handling lines where accumulation and varying goods weights require even speeds. The graph shows how, on a conveyor with self-braking, deforming wheels, speed varies with goods weight.

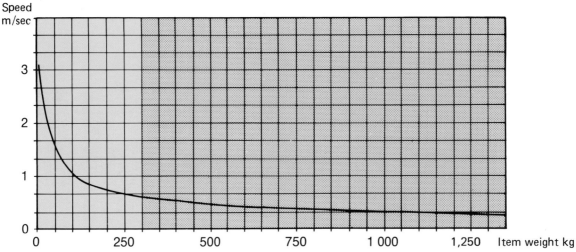

Speed variations according to weight of goods item on a wheel conveyor equipped with deforming self-braking wheels. With items weighing over 300 kg, weight has only minimal effect on speed.

Wheel conveyors can also be designed to flex sideways and to expand and contract for ad hoc routing.

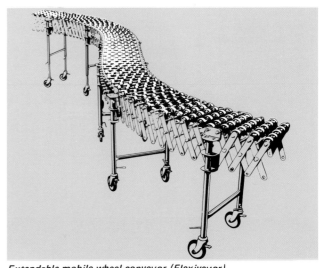

Extendable mobile wheel conveyor (Flexiveyor).

System design

Routing is achieved in roughly the same way as with roller conveyors. Wheel conveyor junctions, however, may be simpler than those for roller conveyors, as the wheels initiating a spur track can be accommodated between those of the main track.

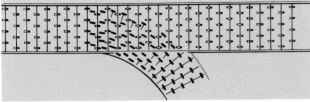

Wheel conveyor junction.

Wheels are often used for transferring goods from a roller conveyor.

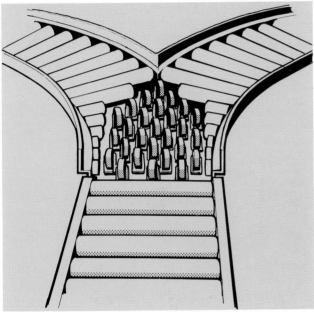

Switching by directional castors.

Directional castors can be used to switch roller conveyors at rates up to 100 cardboard boxes a minute.

Special 'all-way' wheels offer two-way rotation to permit powered and unpowered transfer between lines.

'All-way' wheels.

A height adjustable section of roller conveyor may also incorporate powered wheels between the rollers.

Wheel conveyors interposed between rollers to switch goods from a roller conveyor.

The roller conveyor illustrated is inclined, and goods roll off without powered aids. Wheel conveyor junctions may also be powered for quicker and/or safer switching.

Accumulation occurs automatically with most wheel conveyors, as they are normally unpowered. Powered accumulation, as with roller conveyors, is however possible.

Application examples

Here are a few examples of the way wheel conveyors may be used:

Wheel conveyor moving plastic bins between picking and sorting in a wholesale warehouse.

145

Live storage wheel conveyor installation (a competitor to roller conveyors). Suitable for lightweight cardboard boxes.

Moving pallet loads on a wheel conveyor.

Wheel conveyor for transporting airfreight pallet loads.

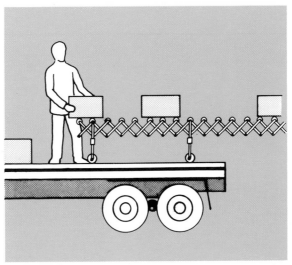

Vehicle loading using an extendable mobile wheel conveyor (Flexiveyor).

EXAMPLE OF WHEEL CONVEYOR SPECIFICATIONS

Conveyor type	Suitable goods	Wheel type	Load per wheel kg	Capacity kg/m	Section length mm	Conveyor width mm	Remarks
Light duty wheels in profiled steel	Cardboard boxes, etc., max 20 kg per item	Plastic	2.5		3,000	n × 90	Suitable for live storage of cardboard boxes, e.g., for picking.
Light duty wheels on steel tubes	Cardboard boxes	Plastic	10–50				
Wheels on shafts		Plastic Steel	5–10 10–50	100–2,000	3,000 3,000	200–600 200–600	
Wheels in side frames	Pallets, timber bundles	Cast iron	200–900	600–4,500	3,000	600	Goods moved directly on the conveyor or in inverted channels.

Belt conveyors

A belt conveyor consists of an endless belt passing over rollers and forming a continuous moving surface which supports goods.

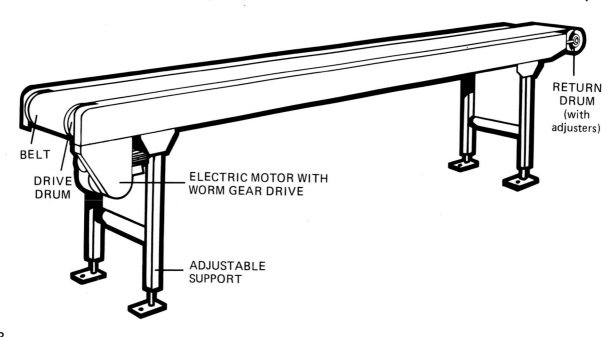

BELT

DRIVE DRUM

ELECTRIC MOTOR WITH WORM GEAR DRIVE

ADJUSTABLE SUPPORT

RETURN DRUM (with adjusters)

Conveyor belts are made from several materials:
- Rubber
- Fabric or plastic
- Steel
- Steel mesh
- Combinations of the above

Rubber belt conveyors are the most common.

Application

Both belt and slat conveyors (see separate section) differ from roller and wheel conveyors in that the carrying surface (belt or slats) accompanies the goods. Thus, in most instances these conveyors *exist as powered units only*.

Belt conveyors mainly handle bulk goods and, to a lesser extent, comparatively lightweight items. Capacity is comparatively high (high speeds are attainable) and many types of goods and packages (bags, small parcels, cardboard boxes, sacks, wooden boxes) can be handled.

Belt conveyors handle goods gently and have the considerable advantage of being able to manage inclines of at least 15°–20°. Tread patterns and flights, which increase friction, permit inclines up to 40°.

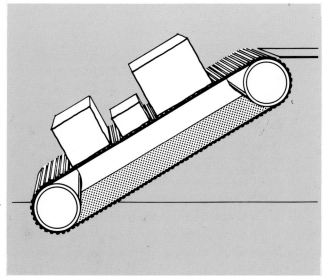

If special tread patterns are used to increase friction, belt conveyors can manage inclines up to 40°.

For this reason, belt conveyors are often used in storage installations where goods have to be moved between different levels. They are then an alternative to elevators of various types (see section 'Elevators'). They demand more space than elevators but may provide higher throughputs.

Loading/unloading lorries, demountable bodies of aircraft are other applications where their ability to bridge different levels is valuable.

The major advantage is the difficulty or impossibility of achieving accumulation, as this would demand low friction between goods and line and would cause goods to slide. (Exception: steel belts.)

This combination of characteristics means belt conveyors are often used with other types of conveyor, such as wheel and roller conveyors. They are installed on inclines and other locations where accumulation is not required; wheel or roller conveyors complement them for accumulation purposes.

For horizontal transport of parcels, boxes of different kinds and other mixed lightweight loads, belt conveyors may even be an alternative to roller and wheel conveyors where there are high throughputs and accumulation is not necessary. If the nature of the goods or the capacity demand belt conveyors, the choice may be a question of price. A belt conveyor may well prove the less costly solution as it has fewer rotating parts and a simpler drive.

Belt conveyors are less attractive for complex sorting and distribution systems as goods normally cannot accumulate (except on steel belts). In addition, designing efficient and safe transfers and junctions is more difficult. Such transport systems, therefore, frequently combine belt conveyors with roller and/or wheel conveyors.

Another drawback may be that the drive pulley and belt return require so much space that the structure will be taller than an equivalent roller conveyor. Also, the belt must be tensioned to a fairly constant value, demanding repeated adjustments unless automatic tensioning is incorporated.

Steel belt conveyors are used primarily for heavy bulk goods although another important application is in the food industry where hygiene is essential (e.g., meat packing plants, bakeries, dairies, etc.). Other steel belt applications include environments which would degrade normal rubber belts — for instance, conveying oily, hot or abrasive goods. Lately, steel belts have found a market in installations for sorting parcels, cardboard boxes and bins when the belt itself acts as an address carrier. (See also section 'Application examples'.)

Belt conveyor characteristics	
Advantages	Disadvantages
Gentle goods handling. Inclined transport. High throughput High speed Low noise level. Adaptability to different types of goods.	Accumulation difficult. Complicated marshalling. Rubber belts affected by hot and sharp materials

Design

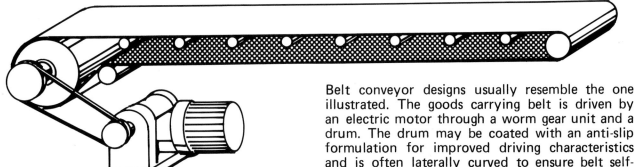

Belt conveyor designs usually resemble the one illustrated. The goods carrying belt is driven by an electric motor through a worm gear unit and a drum. The drum may be coated with an anti-slip formulation for improved driving characteristics and is often laterally curved to ensure belt self-centering. The belt passes over idlers which support and also guide the belt. Idler distances depend on the weight of goods for which the conveyor was designed. For lightweight goods (max. 40 kg) idlers may be replaced by a slider plate. Sometimes the plate is complemented by a few idlers permitting conveyor lengths up to 80 metres with low loads.

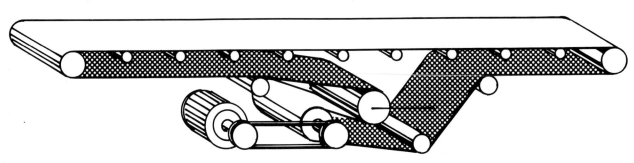

The drive system described makes great demands on space at the end of the line. As an alternative, it may be located centrally under the conveyor (see figure) although this is more complicated and consequently more expensive.

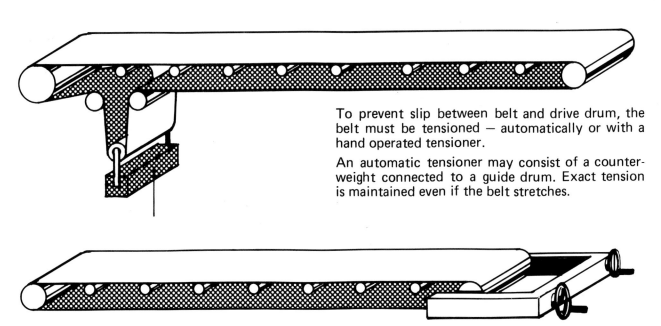

To prevent slip between belt and drive drum, the belt must be tensioned — automatically or with a hand operated tensioner.

An automatic tensioner may consist of a counter-weight connected to a guide drum. Exact tension is maintained even if the belt stretches.

Short conveyors have a fixed tensioning arrangement (see picture) in which the position of the return drum is adjusted with handwheel screws.

Rubber is most commonly used for belts conveying lightweight goods such as boxes and bins and some bulk goods. Reinforcement is usually provided by synthetic material (polyester and polyamide) and a wide range of qualities is available for different duties. Some belts can accept hot materials at temperatures exceeding 90–220°C. Some are oil resistant (chloroprene or nitrile rubber which does not swell from interaction with oil); some convey food and some are fire resistant.

Belt surfaces depend on the type of goods carried and the inclines encountered. Smooth rubber belts are suitable for angles of up to 15° when conveying boxes and similar items.

For inclines over 15°, special friction patterned belts or belts with flights may be used, the latter type mainly for bulk materials. Friction patterned belts permit the following maximum angles of inclination:

Goods type	Maximum angle in degrees
Sacks	30–35
Wooden boxes	35–40
Cardboard boxes	30–35
Beer crates	25–35

Steel mesh belts are suitable for hot and cold materials and to allow drainage from wet materials.

Articulated steel mesh conveyor.

Polyurethane O-belts allow powder and granular residue to drop through.

O-belt conveyor.

For conveying goods up very steep inclines pairs of belts hold goods between them by pressure (see figure).

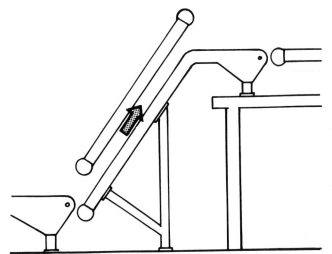
Double-belt conveyor uses inter-belt pressure to handle goods.

Fabrics and plastics are used primarily to convey lightweight materials. Stainless steel is chosen for reasons of hygiene in the food industry and for sorting installations where the belt carries a magnetic code (see section 'Application examples').

System design

Changes from inclined to horizontal conveyor movement should be gradual and smooth and should be completed before the next, horizontal conveyor section takes over.

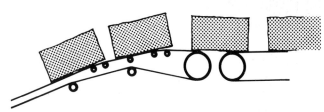

In changes from horizontal to inclined movement the junction must occur where the horizontal movement ends. In this case changeover is less smooth.

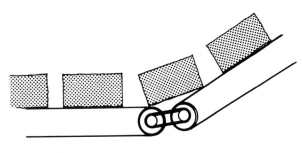

For purely horizontal changes of direction, the belt is supported on tapered rollers throughout special curved sections. Relatively small radii (only slightly larger than belt width) are possible.

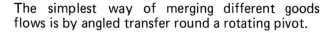

The simplest way of merging different goods flows is by angled transfer round a rotating pivot.

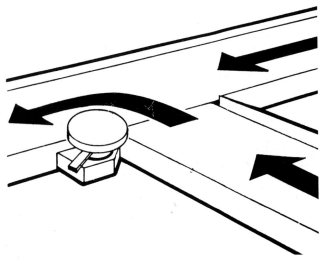

Where conveyors run at different levels, goods flows can be merged via a simple slide.

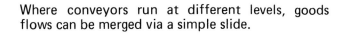

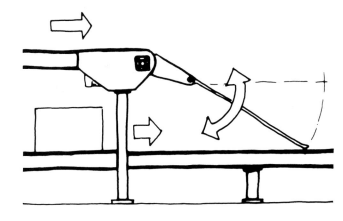

A third method of merging flows is with traffic controllers.

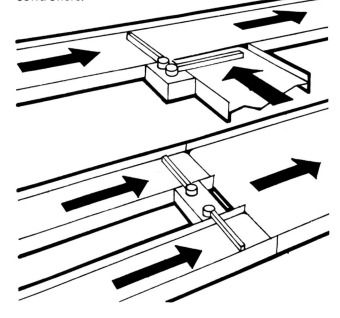

Re-routing can be effected with an electric, pneumatic or hydraulic deflector. This assumes that friction between belt and goods is not so great as to damage the goods during the operation.

Goods may also be redirected with an elevator.

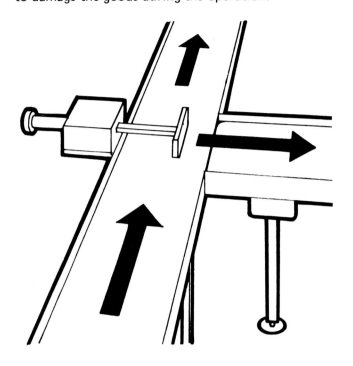

The gentlest re-routing method is to divide the belt conveyor into several sections and to place elevatable chain conveyors in between. This is comparatively complicated and expensive and is mainly appropriate for heavy boxes and bins which cannot be handled in any other way.

Application examples

Conveying bins in a wholesale warehouse

Bins are used in a warehouse to move picked goods to the collation and picking points. Rubber belt conveyors move the bins between the various levels at which picking occurs. This is a good example of moving lightweight items vertically and achieving high throughputs (high conveyor speeds are possible).

Loading and unloading vehicles

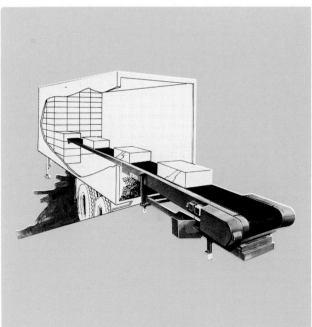

Mobile as well as stationary installations are possible. The picture shows a mobile extendable belt conveyor for lorry loading/unloading.

Horizontal movement of bins in an order picking warehouse

The central depot of a chain store uses a belt conveyor between sorting and a line where goods are loaded onto pallets. Because the belt conveyor engages at both ends with a roller conveyor, accumulation facilities are not necessary. In this case a belt conveyor has proved to be the cheapest and simplest solution relative to throughput requirements and to the weight of goods items.

Steel belt conveyor for sorting parcels

This parcel sorting terminal uses a steel belt conveyor for internal transporting and sorting. Addresses are recorded on the belt in the form of a magnetic spot immediately in front of each item carried. The code corresponds to a switching point where the item is routed to its destination by a deflector. The relatively low friction properties of a steel belt ensure that the deflector does not damage the goods.

Belt conveyor for used cartons

A light duty belt conveyor moves used cartons in an order picking warehouse serving the retail trade. After picking, the cartons are placed on the nearest belt conveyor for transport to a waste compactor.

EXAMPLE OF BELT CONVEYOR SPECIFICATIONS

Conveyor type	Belt material	Power input kW	Max. incline	Speed m/min.	Suitable goods type	Standard width mm	Standard length approx. m	Standard item weight kg	Max. capacity kg/m
Light duty conveyor on slider plate	PVC	0.2	15–25°	15–40	Cardboard boxes, etc.	200–800	3–6*	10–30	40–60
Light to medium duty conveyor belt on idlers	Rubber	1–2	15–35° (greater with friction pattern)	20–40	Cardboard and other boxes, bins	300–800	30**	100	100–200
Medium duty belt on idlers	Rubber	2–3	15–35°	25 approx.	Boxes, bins, pallets	300–800	30**	300	800

*Considerable increase in length possible if idlers are added.
**Considerable increase in length possible wity lighter goods items.

Slat conveyors

Although slat conveyors come in a variety of types all consist of flat or shaped lengths of steel, wood or some other material, mounted on, and following the movement of, chains to form a continuous hard, rigid surface for carrying goods.

Application

Slat conveyors are suitable primarily for moving special materials and goods between fixed locations or stations which are likely to be permanent for some considerable time. Their function is basically similar to that of belt conveyors and they can negotiate inclines where continuous transport without accumulation is acceptable. Unlike belt conveyors they can handle heavy materials.

For instance, they can handle heavy bulk materials unsuitable for belts. They are also used for inclined movement of pallets and other heavy items weighing from a few hundred kilograms up to several tonnes.

On occasion, a slat conveyor may be the only possible solution for conveying goods with sharp edges, feet, etc., unless the shape of the goods is modified with skids or complementary load carriers.

In assembly lines for heavy vehicles and machinery, a slat conveyor may be recessed in the floor. The vehicle wheels may rest directly on the conveyor surface.

Design

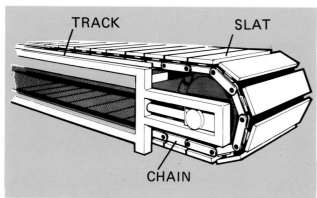

Basic design of slat operator.

Slat conveyors normally have two chains connected by slats. The chains run on tracks with sliding or rolling friction (see section 'Chain conveyors').

Wood is the most commonly used slat material, at least for moving general mixed cargo. It has good friction properties, is cheap and is used to convey, for instance, pallets and heavy mixed goods.

Materials and goods items which are very abrasive or very heavy, require steel slats. These are available in many designs.

The characteristics of slat conveyors are summarized in the table below.

Slat conveyor characteristics	
Advantages	Disadvantages
No friction between goods and conveyor Horizontal and inclined routes Tolerance to heavy loads Replaceable slats Reliability Goods flexibility	Goods accumulation difficult High noise levels Low speeds High power consumption

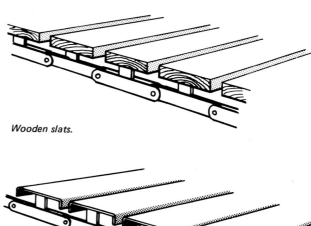

Wooden slats.

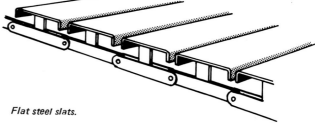

Flat steel slats.

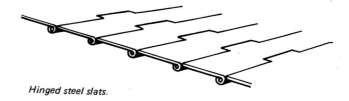

Hinged steel slats.

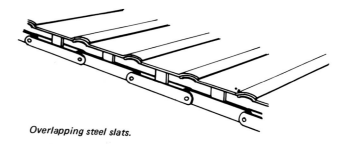

Overlapping steel slats.

Stainless steel slats are used in the food industry for hygienic reasons to convey, for example, fruit and vegetables in bulk between processing stations.

System design

Unlike belt conveyors, slat conveyors may profitably have combined horizontal and inclined sections within the same line.

Neither switching nor accumulation is particularly common with slat conveyors.

For switching purposes, however, slats may be staggered laterally (see figure), a method used in the bakery industry,

Staggered slats may also be used in sorting systems.

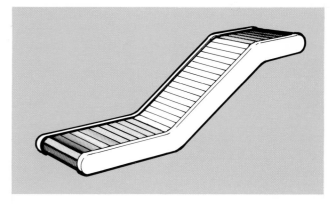

Goose-neck slat conveyor.

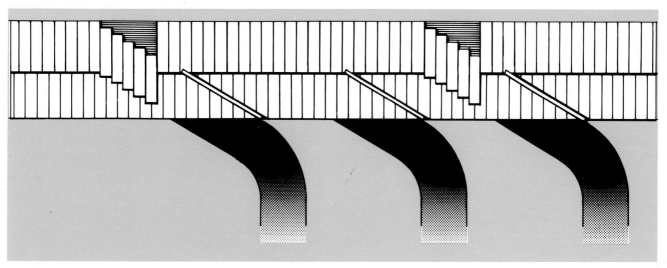

Switching to other lines may be achieved by a combination of staggered slats and fixed deflectors.

Switching with tilting slats is employed in sorting.

A sorting installation with tilting slats. When an item has reached its destination one or several slats (depending on the length of the item) are tilted, and the item slides down to the correct section under its own weight.

Development trends

Slat conveyors are likely to remain popular for handling bulk goods and unpackaged materials (for example, fruit and vegetables) and luggage, (i.e., where the nature of the goods means there is no alternative and there is a continuous heavy flow).

Slat conveyor handling of pallets and similar loads, however, can be expected to decrease as it is replaced by more flexible systems based on trolleys, AGVs, etc.

Application examples

Pallet handling at goods reception

Goods reception at Saab-Scania where pallet loads move up an inclined conveyor based on wooden slats.

Conveyor with steel slats mounted on a loading bay for handling pallet loads. A major advantage is that the operation of trucks and other handling aids is not hindered.

Luggage handling at airport

Conveyor with overlapping rubber slats for handling luggage at airport.

Lorry assembly line

Double slat conveyor set in floor (flat top conveyor). Used in lines for assembly inspection and final adjustment.

EXAMPLES OF SLAT CONVEYOR SPECIFICATIONS

Conveyor type	Goods type	Speed m/min	Capacity kg/m	Standard length m	Width mm
Rubber slat conveyor	Luggage	Up to 100	150	100	300 – 1,000
Light duty wooden slat conveyor	Mixed cargo	8 – 10	20 – 100	60	400 – 800
Medium duty wooden slat conveyor	Pallets	8	1,000	6	1,250 (Euro pallets)
Loading bay conveyor with steel slats	Pallets	5 – 10	1,000 – 3,000	50	900
Vehicle conveyor with steel slats	Vehicles	1 – 10	800 – 6,000 (point load)	100	Varying

Tray conveyors

Tray conveyor as part of an installation for sorting mixed cargo.

With tray conveyors, the trays circulate continually or step-by-step in a closed circuit, usually at bench height. Trays are normally linked by chain.

The major difference compared with belt and slat conveyors is that items are allocated their own trays. As trays can accept address codes, identification, switching and sorting is facilitated.

Application

Assembly operations (e.g., accumulation and distribution of materials and components to assembly stations) are typical applications. Trays are excellent accumulation devices since they can circulate their loads until assembly stations need them.

Tray conveyors are common in the electronics and radio/television industries (see application example below) as well as in sorting systems for mixed cargo which employ the trays' address-carrying facility.

Trays may be tilted either way. Note the difference between tilting trays and tilting slat conveyors. In the latter case, a number of tilted slats is selected according to the length of the load; magnetic addressing is therefore not possible, and switching must be controlled by a central memory monitoring the location of individual items. Tray sorters, however, offer the simpler addressing system, an especial advantage in complex conveyor systems although it requires that tray size is adapted to the largest item carried.

System design

Loading and unloading lightweight goods are operations usually performed by pneumatic or hydraulic deflectors. Heavier goods may require tilt-trays equipped with rollers.

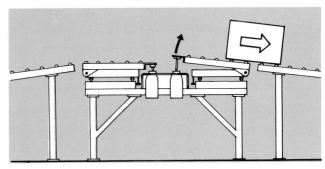

Unloading a conveyor by tilting the trays.

An alternative to tilt-tray loading/unloading may be to equip the trays with powered rollers. This method is used, for example, in sorting installations.

Design

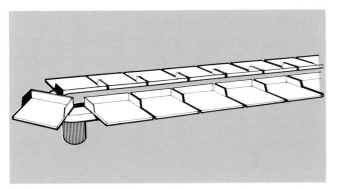

Tray conveyor design principle.

Trays may come with or without raised sides; one or several chains supply the drive. Operation may be continuous, with the trays attached permanently to the chains, or the trays may be added to the moving chains as and when required.

Application examples

Rotating picking store for small items

A multi-tier tray conveyor can serve as a rotating order picking store where the picker is able to reach the entire range of components from one position.

Lightweight goods installations

A sorting installation consisting of a tilt-tray conveyor designed for sorting mail parcels, plastic bags, sacks and similar items weighing up to 20 kg. Sorting throughputs up to 10,000 items per hour can be achieved.

A tray conveyor is used to transport picked orders to the packing station.

Pallet sorting installation

Chain store distribution centre uses a tray conveyor of unconventional design. The trays have individual address codes in the form of mechanical tongues and are equipped with powered rollers for switching pallet loads to connected roller conveyors. The trays travel round a circuit consisting of a powered accumulating roller conveyor.

Tray conveyor specifications

As tray conveyors are fairly rare, it is difficult to speak of standard systems characterized by specific dimensions and performance figures. Installations are mostly custom-built for specific purposes, and little catalogue information is available from manufacturers.

Chain conveyors

Chain conveyors consist of one, two or more chains running in parallel tracks, loads being carried directly on the chains.

Application

Chain conveyors are widely used over both short or long distances, carrying goods which are unsuitable for roller conveyor. As the chains both accommodate and accompany the goods, base requirements of individual items are not very exacting.

Three strand chain conveyor transferring pallets and half-pallets between roller conveyors.

Transverse movement of pallets (with their baseboards placed across the chains), often in connection with roller conveyor systems, is a typical application.

Chain conveyors can then be used for lateral transfer and distribution between different roller conveyor lines. Similarly, they can be used for lateral transfer at pick-and-place stations in mechanized high level storage installations.

Chain conveyors are also used for wooden boxes and bins with uneven undersides. Chain conveyors, possibly supplemented by flights, may be the natural choice for conveying lengthy goods such as timber and steel.

In chemical engineering and the food industry, chain conveyors play an important part. Simple chain conveyors made of plastic or stainless steel handle bottles, beer cans and similar items. (See section 'Application examples'.)

| Chain conveyor characteristics ||
Advantages	Disadvantages
Undemanding goods base requirements Suitable for handling lengthy goods. Horizontal and vertical changes of direction easily arranged. (This does not apply to bowing chain conveyors).	Accumulation hard to achieve. Accumulation which depends on lack of friction between chain and goods requires acceptable goods base and high power inputs. (Plastic chain conveyors for small items are an exception). Generally speaking, sliding chain conveyors require relatively high power inputs. (Plastic chain conveyors are an exception).

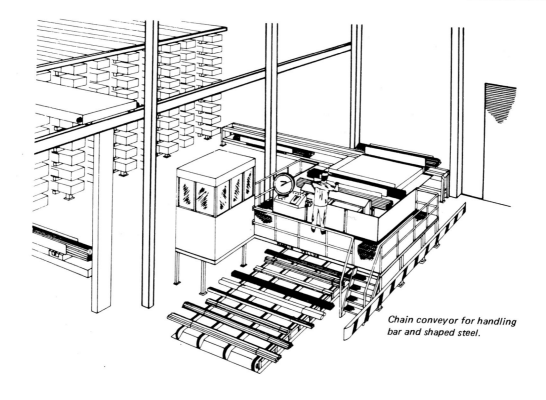

Chain conveyor for handling bar and shaped steel.

Design

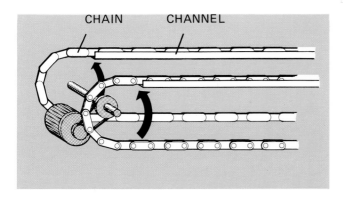

CHAIN   CHANNEL

Bowing chain made of steel

Chain conveyors consist of one or several chains on a frame between a driving wheel and a return wheel. The chains run in channels with sliding or rolling friction.

Flat link chains may or may not incorporate rollers.

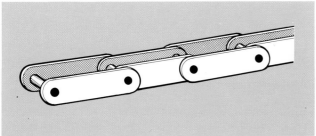

Chain without rollers.

Chains without rollers are used for low friction work, mainly with light and medium loads (up to 1,000 kg).

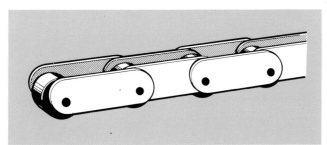

Chain with rollers.

Chains fitted with rollers are used for heavier loads, e.g., for transferring pallets, and usually have raised sideplates to support the loads.

Bowing chains articulate both horizontally and vertically, providing a high degree of layout flexibility, with comparatively acute vertical and horizontal changes of direction. (For minimum radius data, see end of section.)

Bowing chains can be of steel (as in the top right hand picture) or plastic (cellulose acetate or poly-acetal). In either case they are used as sliding chains.

Steel chains transport wooden bins and pallets weighing up to 1,000 kg. The coefficient of friction between chain and track is 0.3; fairly high power inputs are needed for heavy loads.

Friction can be halved if the underside of the chain is lubricated by routing it along a roller set in an oil bath.

Bowing chains have the advantage of being comparatively easy to recess into floors. Floor level handling of pallets, etc., is consequently cheaper than by equivalent roller or slat conveyor. On the other hand, more power is necessary and noise levels are higher.

Bowing chain made of polyacetal.

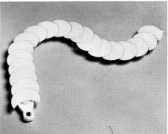

Bowing chain with additional load bearing discs providing a larger contact area. As a result, large items can be handled on a single chain with greater stability.

Plastic bowing chains are available with integral plastic links and stainless steel studs (see picture above). The plastic results in quiet operation and a low coefficient of friction (0.05 lubricated, 0.15 with no lubrication). It is also corrosion-proof and endures operating temperatures of $-40°$ to $+80°C$ (-40 to $+175°F$), making this type of conveyor chain useful where hygienic requirements are high — e.g., breweries, dairies and restaurants (see section 'Application examples').

Curve radii down to 130 mm are possible. Plastic chains are primarily suited to lightweight goods. Items must not weigh over 200 kg.

Bowing chains can be given additional bearing discs (see picture). A similar type of chain, in stainless steel, is for applications where wear and heavy loads disqualify plastic materials.

System design

Direction changes — switching

Chain conveyors with conventional steel links for heavy goods run in straight lines only and, for the most part, horizontally. Bowing chains, however, can follow bends, and the track channels are shaped accordingly (see picture).

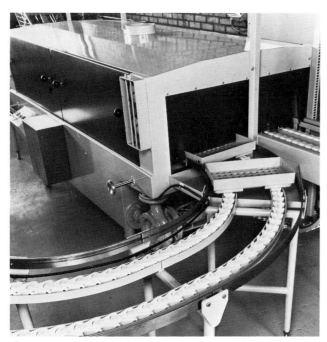

Bowing chains can follow a curving track.

Note, however, that goods will twist on the bends if the chains are allowed to run at the same speed. To avoid this a freely-rotating turntable may be installed to cover the inner chain at the bend (see picture).

A freely-rotating turntable covers the inner chain at the bend to prevent load twisting.

Drawing two conventional chain conveyors together at a changeover point is difficult and therefore rare. But chain conveyors themselves are often used as switching devices in roller conveyor installations (see section 'Roller conveyors'). Switching is easier with bowing chains.

With single plastic chain conveyors several chains can be run side by side and goods moved from chain to chain by guides.

Accumulation

Simple accumulation with a conventional steel chain conveyor requires the chains to slip under the loads. Where the coefficient of friction is less than 0.3, this wastes power and is noisy. Instead, a special type of accumulating chain with raised rollers is used (see figure).

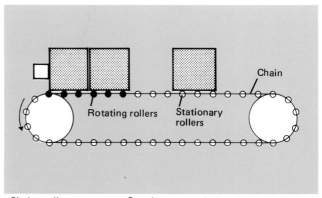

Chain roller conveyor. Permits accumulation by letting rollers under static loads rotate.

Goods are carried on rollers mounted either between strands of chain or directly on chains (as shown in figure). While goods are moving, the rollers do not rotate. During accumulation (i.e., under pressure) the rollers under the goods rotate, allowing the chain to maintain its speed. This assumes that loads have smooth steel undersides or steel skids; in turn this means that the basic advantage of a chain conveyor (i.e., its ability to handle 'difficult' goods) is lost.

If it is desirable to accumulate heavy wooden pallets or boxes without an intermediate bearer or without modification, other solutions will have to be found. Pop-up stations that lift goods, or con-

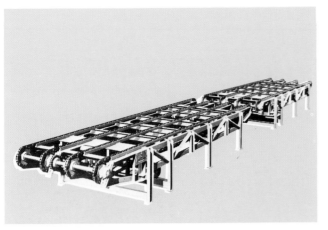

Chain conveyor divided into sections to allow accumulation.

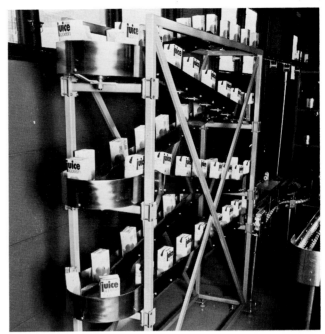

A spiral plastic chain conveyor used for accumulating orange juice packs.

veyor lines that divide into sections (see picture) are two examples.

When moving light loads on plastic bowing conveyors (polyacetal or cellulose acetate), low friction accumulation is possible. The same can be said for bowing chains, but power will have to be increased and noise levels will rise.

Application examples

Moving drink crates

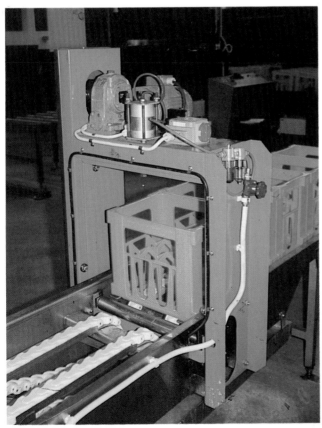

A two-strand plastic chain conveyor mounted in a stainless steel channel, used to convey plastic crates. The installation incorporates combined air operated drive and stopping.

Conveying Euro pallets into an order picking warehouse

A handling installation for Euro pallets consisting of a 120 metre chain conveyor with six built-in roller junctions. The line conveys pallets placed with their long sides across the chains. The transfer line for each pallet is addressed at the time of input. On arrival at its station, the pallet is switched and moves under its own weight down one of 12 roller conveyors extending into the shelving. The system is used for input and distribution of goods to a large shelf-based warehouse.

Transferring pallet loads from hand pallet truck to roller conveyor

One method of handling pallet loads from a hand pallet truck to a roller conveyor which stands more than a foot above floor level is to use a pallet lifting chain conveyor. This may be a cheaper and more flexible solution than a floor-recessed lifting table. The chain conveyor has three chains, one for each baseboard. A few hydraulic strokes moves the pallet onto the conveyor which then elevates the pallet. The operation takes 3—4 seconds.

Feeding scrapped cars to a press

Heavy chain transporters can handle loads of up to 5 tonnes. This conveyor feeds scrapped cars to a press and provides an example of how a chain conveyor can provide a fairly simple and cheap method of moving cumbersome loads if accumulation is unnecessary.

EXAMPLES OF CHAIN CONVEYOR SPECIFICATIONS

Conveyor type	Chain material	Friction coefficient (chain line)	Suitable goods	Item weight kg	Goods width for single chain mm	Max. chain load (tensile stress) kg	Speed m/min.	Min. curve radius mm
Bowing chain conveyor	Plastic (polyacetal, cellulose acetate)	0.05 (lub.) 0.15 (unlub.)	Consumer packages, crates, bottles, tins	0—200	50—120	200	1—35	130
Bowing chain conveyor*	Steel	0.15 (lub.) 0.3 (unlub.)	Bins, pallets	0—1,000				250
Medium duty conventional chain conveyor	Steel		Euro pallets, half-pallets, other pallets, wooden boxes	1,000			5	Normally no bowing
Heavy duty chain conveyor	Steel		Lengthy goods, timber bundles steel	5,000			5	Normally no bowing

*Dogs used for inclines exceeding 12%.

Overhead conveyors

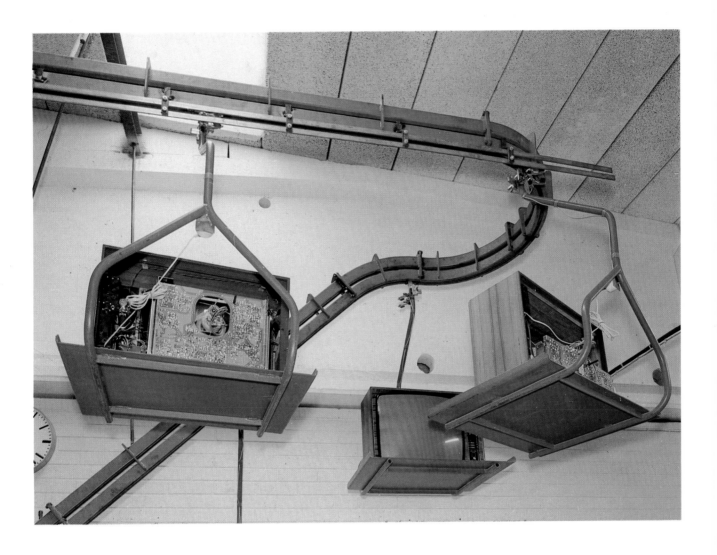

'Overhead conveyors' describes all conveyors with roof-mounted drives. Literally speaking, loads can be transported by carriers running on rails suspended from the roof or carried on the floor by vehicles with roof-level drives.

Overhead conveyors may be divided into three main types, depending on the drive method (see figure). The usual method is by chain. Worm drives and self-powered carriers are two other options.

Chain operated overhead conveyors come in two main designs: continuous direct drive or with a disengaging feature via a separate chain fitted with drive dogs (power-and-free type) to facilitate accumulation, switching, etc.

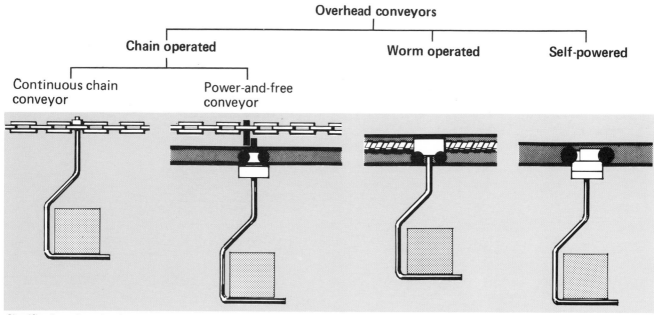

Classification of overhead conveyors.

Application

Overhead conveyors provide a fixed handling installation that leaves the floor completely free for personnel and vehicle traffic. Great freedom of layout is possible. Both vertical and horizontal bends are easily constructed. Lines can accommodate the obstacles encountered in a non-purpose-designed building.

Overhead conveyors are often comparatively cheap and simple. Because of their favourable weight/handling capacity ratio, overhead conveyors can often provide both a comparatively inexpensive and simple solution to a handling problem. Capacities can be adjusted by allowing several carriers to handle a load. An overhead conveyor designed for 100 kg items is able to handle loads up to 400 kg suspended from spreader attachments. Another advantage — extensively used in factories — is the simplicity of accumulation.

On the negative side, an installation may limit the usable free headroom. This could be critical in warehouses with high lift fork trucks. The load bearing capacity of the roof also becomes important; if reinforcement is necessary, considerable extra costs may be involved.

The most extensive use of overhead conveyors is in industry, particularly in assembly operations where both basic assemblies and components are moved in a way that leaves the floor area free for traffic, etc.

For certain operations such as paint spraying and coating, overhead conveyors are the only possible method. Blasting and enamelling installations also employ overhead conveyors to overcome difficult operating conditions.

Goods terminals (for combined rail/road operations as well as for road vehicles alone) use overhead conveyors for internal transport and for sorting mixed general cargo. In these cases, hand pallet trucks are towed along the floor by suspended hooks. The mixed handling (by manual means, hand pallet and other trucks) at these terminals demands floor areas free from obstacles, and overhead conveyors are the only realistic solution.

Overhead conveyor at a goods terminal.

Warehouses and stores can use overhead conveyors both for input and output, e.g., between shelving. The advantage is that they do not hinder trolleys, trucks and personnel in the aisles.

Pick-and-place installations are now tending to be supplied with packing materials by overhead conveyor. Generally speaking this solution creates the least disturbance for other handling arrangements. Occasionally, this requirement is entirely forgotten at the planning stage; an overhead conveyor is then a convenient answer.

In addition to mixed cargo and assembly operations there are also several types of installation in which the loads themselves are more naturally transported and stored in a suspended position, e.g., clothing, animal carcasses. Development in such areas is towards transport and handling lines which use overhead conveyors wherever possible.

Overhead conveyor for clothes on hangers.

Summary of overhead conveyor characteristics	
Advantages	Disadvantages
Free floor area	Limited headroom
Layout simplicity (vertical and horizontal bends)	Roof must meet certain load-bearing requirements
Simple design	
Low investment	
Accumulation possible (power-and-free)	

Design

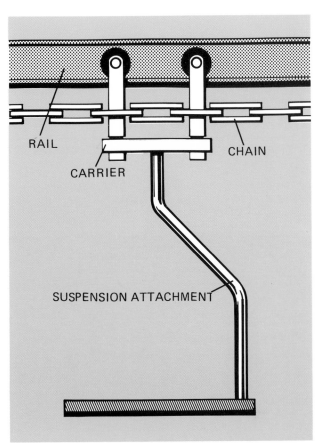

RAIL

CARRIER

CHAIN

SUSPENSION ATTACHMENT

The figure shows the design principles of a chain operated overhead conveyor. Loads are suspended from wheel-mounted **load carriers** which are moved on a **rail** by a chain. In this case, the chain is beneath the rail.

A complete conveyor installation may incorporate the following components:

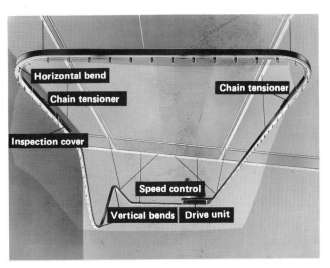

Horizontal bend

Chain tensioner

Chain tensioner

Inspection cover

Speed control

Vertical bends   Drive unit

171

Rails and chains come in many different designs.

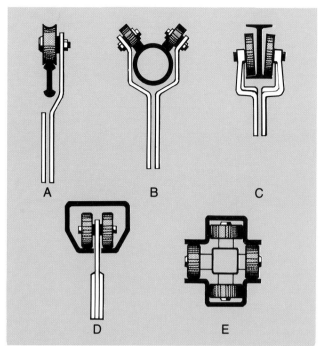

Various rail/chain designs. A, B and C represent alternatives with carriers mounted on external rail surfaces. In D and E both carrier and chain run inside the rail.

Chains which operate inside the rail normally have designs which permit horizontal and vertical articulation, together with runners which provide a type of **bowing action.**

Bowing chain.

This is the simplest design to install. It also protects carrier wheels and bearings.

Suspension attachments can always be adapted to the type of goods carried. Many versions exist. Their design may be influenced by the need to load and unload, and to switch or rotate the loads.

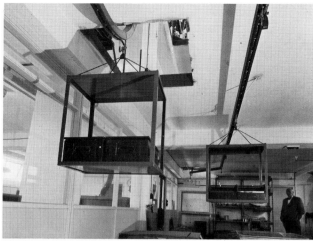

Suspended platforms used to carry mixed cargo.

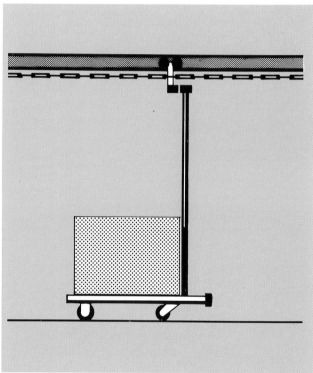

Floor trolleys are used in, among other places, mixed cargo terminals.

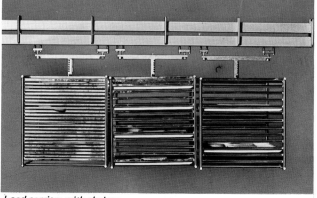

Load carriers with shelves.

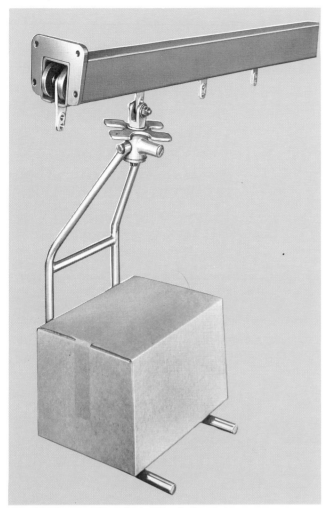

Load carrier consisting of a two-tine fork.

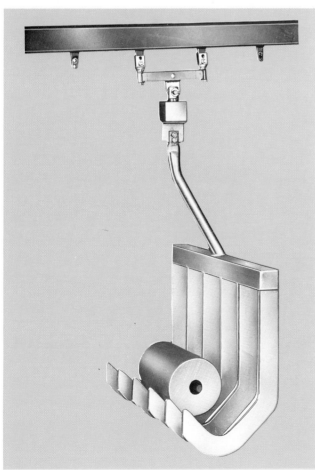

Finger hook.

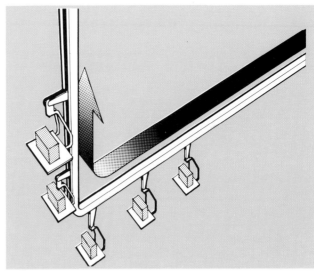

Articulated suspension attachments facilitate vertical handling.

Conveyed objects may require rotating, for instance, in spray painting. Special attachments are available for this.

These are a few examples:

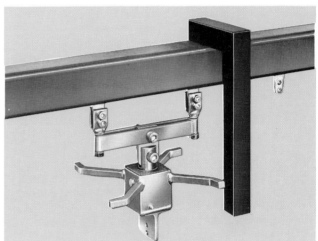

Automatic rotating attachment: fixed positions.

Automatic rotating attachment: infinitely variable positions.

Power-and-free accumulating overhead conveyors

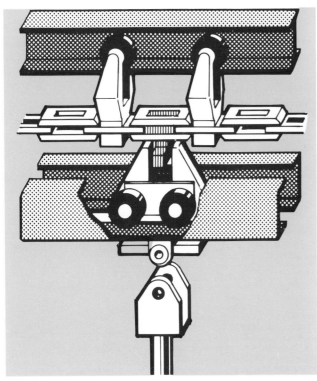

If individual goods items have to be released, for example, in connection with accumulation or switching, two running rails are used. The continuously moving drive chain runs on the top rail, while the bottom rail accommodates separate carriers driven by pusher dogs. The name 'power-and-free' thus refers to the release facility.

Of the two release methods, the simpler uses spring-loaded followers. When a carrier is stopped by a stationary carrier ahead, the spring-loaded follower gives way to let the drive chain pusher dog pass. This simple and inexpensive accumulation is limited, however, by the fact that maximum tractive power is determined by the spring strength of the follower, making inclined or vertical handling difficult.

A more complex but quite commonly used method is based on mechanical release of a vertically operating follower (in a so-called proximity carrier).

Power-and-free conveyors offer simplified accumulation and switching and can also be used when the route involves a brief descent to a local level.

They are commonly applied in the manufacturing industry although, latterly, self-powered overhead monorail systems (see below) have become more competitive because of the greater flexibility inherent in individual drive systems.

To save drive chain costs, power-and-free conveyor systems can be designed so that an inclined bottom rail (2—3 per cent) in some sections moves the trolleys. The technique is used for accumulation. Note, however, that the carriers may gather considerable speed; only robust goods (such as sacks) should therefore be conveyed in this way.

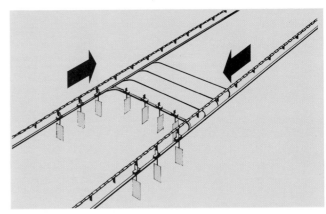

Combination of powered and unpowered sections in a power-and-free overhead conveyor.

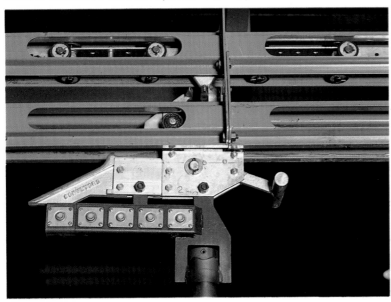

Power-and-free conveyor.

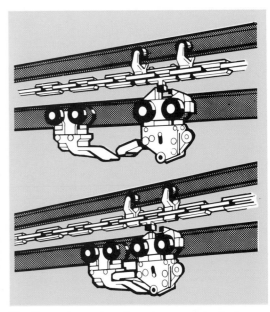

The figure shows a version where a carrier in front releases an approaching carrier by forcing its projecting release handle upwards. Note that the total length of the carrier must always be greater than that of the loads handled.

174

Worm operated overhead conveyors

Substituting a worm for a chain results in an ingeniously simple overhead conveyor. Loads can be suspended on hooks mating with the worm thread. Among drawbacks are comparatively low speed and mechanical wear of the worm, especially if heavy loads are carried.

A similar function is achieved if, instead of a worm, a rotating tube is used and goods are suspended from carriers running on skewed wheels. (Cf. the Car-Trac system described in section 'Rail mounted trolleys, trucks and tractors'.) The design is used, for example, for moving clothes on hangers.

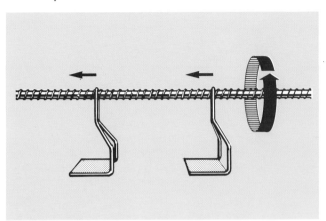

Worm operated conveyor.

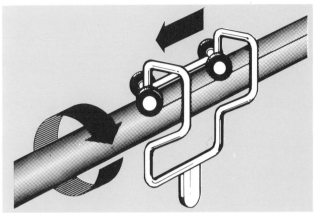

Rotating tube conveyor with skewed carrier wheels.

Self-powered overhead monorail conveyors

Self-powered overhead monorail conveyors have carriers with built-in electric motors instead of chain, worm or rotating tube drives.

Self-propelled carriers can run inside girder structures (as shown in figure). In this case power unit, load carrier and control electronics form a single assembly. Power supply is through contact between a current pick-up on the carrier and a conductor rail inside the girders. This design facilitates routing and control of individual carriers. (See section 'General information on line restricted materials handling equipment'.) Track systems and switching are simpler.

On the other hand, individual carriers are more expensive and more complex, although this may not prevent a system with even a moderately large

number of carriers from ultimately emerging as a less costly solution

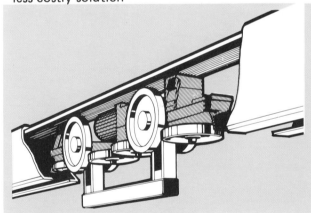

Self-powered overhead monorail conveyor.

Self-powered carriers can also be designed to run externally on girders. They are then more accessible for servicing and can, as shown in the figure, be constructed from a power supply and propulsion and control modules.

Externally mounted carrier.

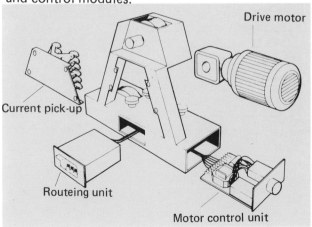

Load-carrier constructed from modules.

System design

The immediately preceding section, headed 'Design', carried an illustration of a possible version of a complete overhead conveyor system.

In many applications for lightweight goods, **loading/off-loading** is entirely by hand. Mechanizing this function is less simple than with other conveyor types.

One way is to use a combination of roller conveyor and lifting table. When an overhead load carrier approaches the loading station, the lifting table on which the goods rest is raised, transferring them to the carrier's fork or hook.

In a similar loading arrangement, a hydraulic or pneumatic ram lifts goods to the correct height at the correct moment (see figure).

Alternatively, the conveyor lowers the carriers which then engage the goods with their hooks. This may require more space as conveyors normally should not be subjected to inclines of more than 30°.

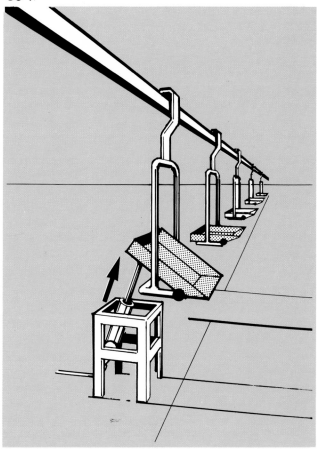

Offloading by tilting the goods carriers.

Automatic offloading of lightweight goods is simply achieved with goods carriers which, for example, are tilted by hydraulic cylinder (see figure).

Hydraulic ram lifts goods for loading.

Switching is possible with overhead power-and-free chain conveyors and self-propelled monorail conveyors but not with simple chain operated overhead conveyors.

Switching devices for power-and-free conveyors are fairly complicated (see figure).

Switch in power-and-free conveyor.

Separate transfer chains (so called caterpillar chains) normally handle transport through the curved junction. Resetting may be electrical, pneumatic or hydraulic.

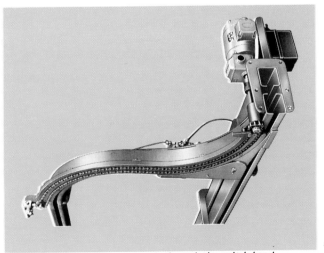

Transfer chain handles movements through the switch bend.

With **self-propelled monorail conveyors**, switching is easier and it is sufficient to move the rail mechanically.

The switching method used can be a simple movement from track to track (see figure).

Switching a self-powered overhead conveyor by moving from one track segment to another.

Switching is also possible by a 'turntable'.

'Turntable' switching.

Application examples

Conveyor system for clothing

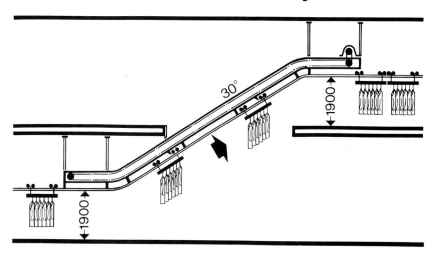

The figure shows a conveyor installation in a warehouse where hung clothes are stored. To move goods between different storeys, a chain operated overhead conveyor is used, with free load-carrying carriers propelled by pusher dogs. Unpowered trolleys handle operations restricted to separate floors where slightly inclined tracks provide movement by taking advantage of gravity. This is a commonly used method of storing suspended goods in distribution warehouses.

Assembly line in electric radiator factory

A chain operated power-and-free conveyor handles components on carriers from a loading station to four assembly stations. The carriers have adjustable routeing pegs set by the operator at the loading station. The assembly station input switches read the address combinations. Routeing information is automatically cancelled by the assembly station.

Sorting bays at mail terminal

The system, based on a chain-operated power-and-free conveyor, delivers mailbags to containers and to hoppers; operations include short-term accumulation. Mailbags are gripped by holders suspended from the carriers and equipped with automatic release.

Each of the 26 stations triggers the automatic release and receives sacks into a mail container. Routeing is based on the analog memory principle (see section 'General information on line restricted materials handling equipment') with a shift register keeping track of the distance travelled by each mailbag relative to its intended destination, and eventually triggering the solenoid which opens the sack holder over the correct mail container.

Self-powered monorail conveyor for transport and sorting at mixed cargo terminal

On its towbar each carrier has a control box which can be coded for the receiving station. Following a pulsed start signal, carriers pass from sidings and on to the main track as soon as it is vacant. They then continue until they reach their respective destinations where they are switched out again. Power supply is by conductor rail and carrier-mounted current pick-ups. Running parallel to the conductor rail is a signal track which communicates with the central control system. The conveyor system switches have two moving parts: blades which guide the carrier drive and steer wheels, and contacts for current and signal. The system is computer controlled. Carrier speed is 20 to 45 m/min., allowing up to seven carriers a minute to pass along a system section.

EXAMPLES OF OVERHEAD CONVEYOR SPECIFICATIONS

Conveyor type	Load per carrier kg	Speed m/min	Min. curve radius mm		Accumulation
			Horizontal	Vertical	
Continuous chain operated overhead conveyor for light goods	25–100*	10–20	500	300	No
Chain operated overhead conveyor (power-and-free) for light goods	25–100	15–60*	300	500	Yes
Chain conveyor (power-and-free) for heavy goods	1,500	10–15	—	—	Yes
Self-powered monorail conveyor, loads towed on floor	1,500	20–50	2,000	3,000	Yes
Self-powered monorail conveyor, suspended loads	500	Normal 10–20 Max. 100	500	—	Yes
	1,500	Normal 10 Max. 80	1,000	—	Yes

*At high speeds loading/offloading may present problems.

Rail mounted trolleys, trucks and tractors

In addition to roller conveyors, belt conveyors, etc., line restricted materials handling equipment includes programmed trolleys and trucks. A distinction is made between two groups, i.e., **rail mounted units** and **automatically guided vehicles (AGVs).** The former are mechanically driven and controlled on fixed rails, while the latter consist of floor-level vehicles controlled by the magnetic field from a conductor wire buried in the floor (see section 'Automatically guided vehicles').

In turn, rail-mounted units can be divided into sub-groups on the basis of using roof- or floor-mounted rails. Systems with common drive or individually powered vehicles are to be found in each group. Carriers are designed either to handle suspended loads or to tow trolleys on the floor.

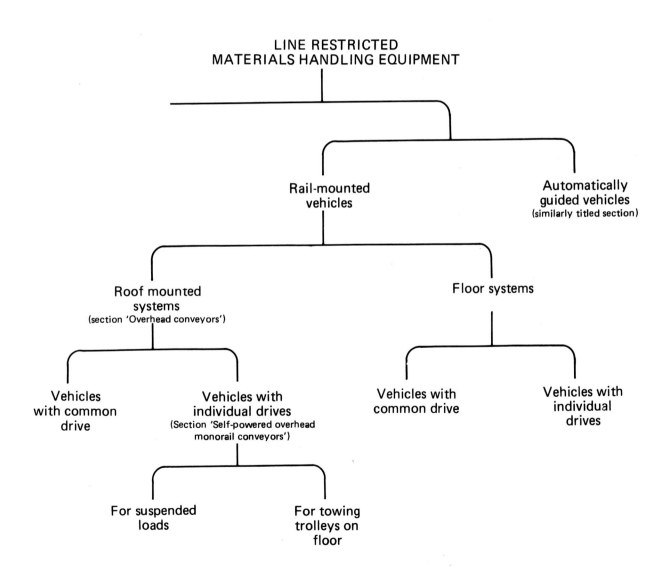

Towline systems

This category primarily includes systems in which vehicles are moved by floor recessed chains. Each vehicle has at the front a vertically adjustable follower which links the chain drive to the vehicle. Relocatable pins carry the vehicle's routeing code.

The switch point has a movable tongue which steers the follower onto a side track with its separate drive. Switching is initiated by pins, positioned to correspond to destinations, actuating microswitches in the floor. The vehicle then enters the siding and stops. When given a new task, the vehicle is returned manually by the operator to the main circuit after having a new destination imposed. In large systems with many vehicles, routing pins are supplemented by bar-code labels to inform an overriding control computer of the position of each individual vehicle.

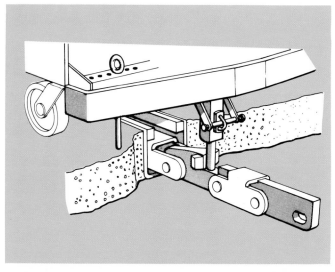

Towline system with chain drive through vehicle follower. A relocatable routeing pin is shown on the left.

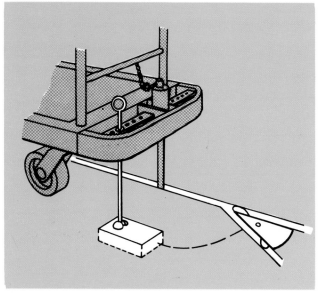

Rod for mechanical track switching.

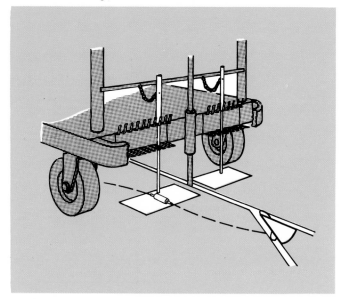

Magnetic switches enable contactless actuation.

In routeing chain-drawn vehicles with a relocatable rod or pin, the latter either mechanically actuates an electric switch which in turn triggers a spring-loaded guide tongue, or else it actuates a magnetic switch which ultimately switches the track electrically.

Normal speed in systems with no track switches may be 45 m/min; this is reduced to 30 m/min where there is switching. Chain pitch can be 100 mm and breaking strain 1.3 tonne, while the tracks may be 75 mm deep and 55 mm wide (external size).

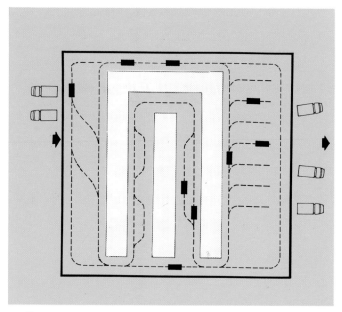

Example of track layout for chain-drawn vehicles.

The common drive principle also applies to vehicles in **Car-Trac systems,** a design from the mid-1960s. The vehicles consist of a frame with four horizontally mounted wheels running against rails. Because the wheel axles are vertical, the configuration provides support and lateral guiding.

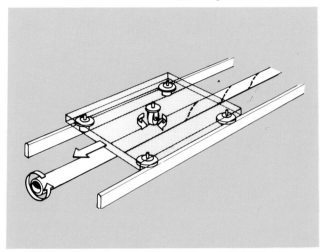

Design principle of Car-Trac: four wheels on vertical axles, with drive transmitted from a rotating cylinder to an adjustable drive wheel on the vehicle.

By changing the set angle of the drive wheel, vehicle speeds may be varied up to a maximum of 150 m/min. Using remotely controlled guidebars, wheels can be set from a central control console. Systems are normally designed to have powered roller conveyor sections where loads are carried. Car-Trac was invented in Sweden but, because of its high price, the system is not widely used there in spite of its excellent technical features. Lately, however, it has been discovered by industry in other countries and is now used in several auto-

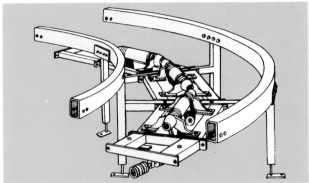

Bend section with sectional drive cylinder.

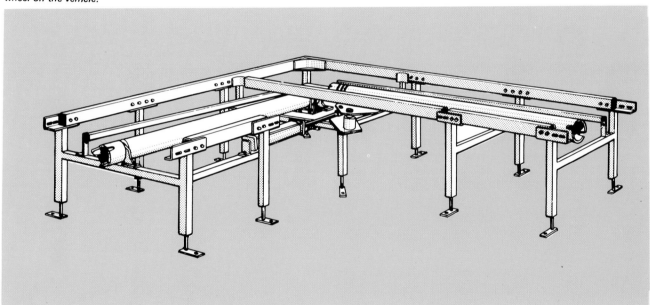

This 90° turn in the track changes the vehicle's travel direction through 90°

The wheel arrangement means that vehicles will move through a right-angle if part of the track is turned through 90°. Thus vehicles and their loads still face in the direction of travel despite the turn. Conventional bends are also available.

Car-Trac vehicles are operated by a rotating cylinder positioned between the rails. Motion is transmitted via a drive wheel attached to the underside of the vehicle, producing a version of the worm transmission.

mated warehouses serving the Japanese automotive industry.

Another type of trolley, **Eli,** with similar characteristics has a frame with four conventional rail wheels and a friction clutch connected to a drive chain between the rails. The clutch permits accumulation of trolleys. The frame is designed so that trolleys may be switched from rails to roller-ways and thus permit intermediate storage, e.g., in a multi-storey live storage installation served by crane. 90° turns are made with turntables and the system will function in an inclined position, thus ensuring convenient working heights at all times. Speed is approximately 30 m/min, and trolleys can be equipped with box bodies to suit different product requirements.

The Swiss **Rollax** design may also be included in this category. It has simple, wheel mounted frames for pallet loads. The trolleys are designed primarily for use in throughflow installations where they replace conventional roller conveyors which, requiring braking rollers, represent a fairly costly solution. Rollax trolleys are mounted on a slightly inclined track where gravity operation is supplemented by an auxiliary drive in the form of a steel cable between the rails. Their design facilitates handling by stacker crane, and they can also remain with their pallets on distribution vehicles. This application offers quick loading/unloading potential and reduced vehicle turnround times.

Similar to Rollax are the French **Mills-K** and the U.S. **Interlake** systems. Pallets are placed on load-carrying frames without wheels and moved by a remotely controlled, rail mounted shuttle car which collects them one by one. Handling throughputs are low but this may be acceptable in an actual warehouse.

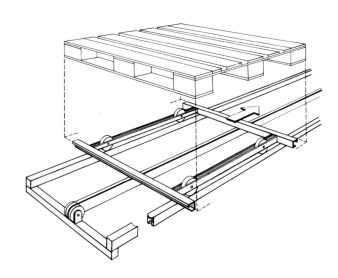

The Rollax system consists of a simple load-carrying trolley mounted on slightly inclined rails. Gravity operation is combined with a steel cable drive between the rails.

Floor systems with individual drives

This group includes rail mounted trolleys normally moved by hand along straight tracks and through right-angle switches. A British make, **Square One**, has trolleys and rails of formed sheet metal. Depressing a pedal under the trolley at special switching stations turns the four wheels through 90°, enabling the trolley to move at right angles to its former direction of travel. The trolleys are pushed by hand, allowing loads of up to 1.5 tonnes to be handled. The system may be useful for moving goods round machinery or for high density marshalling.

A similar design develops the concept further and provides each trolley with a battery powered electric motor. A marshalling (storage) system with a number of parallel parking tracks serviced from a central aisle and designed for Ford in Dagenham during the 1960s may also be mentioned in this context.

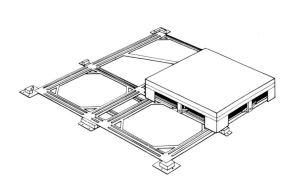

A Square-One trolley consists of a single frame on wheels which can be turned through 90° to change travel direction.

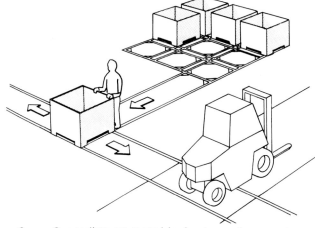

Square-One trolleys are moved by hand on rails recessed into the floor. The rails allow unimpeded truck traffic.

Application examples

Rail mounted trolleys are simple to construct, operate and control. But the (often recessed) rails hinder layout re-organisation, rendering them less suitable for highly flexible applications. Also, the 1980s are likely to see rail mounted trolleys being superseded more and more by AGVs for pure transport operations. Car-Trac and Rollax trolleys, however, will remain in use for special applications.

Application and development

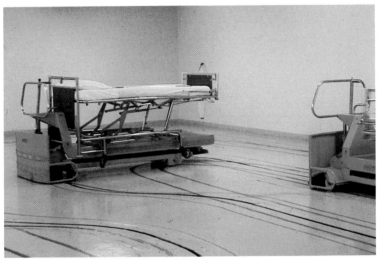

Towline floor trolleys move hospital beds through connecting passage.

The same trolleys can also be used to transport food trolleys.

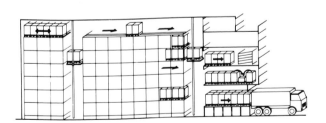

A Rollax equipped warehouse combining accumulation storage (left) and live storage. Service is by stacker cranes; output to multi-storey building for picking and dispatch.

Rail mounted trolley system for luggage handling at air terminal.

Automatically guided vehicles

Line restricted materials handling equipment includes automatically guided vehicles, i.e., automatically controlled driverless trucks. AGVs may operate as tow tractors or as load carrying vehicles. In both cases, they are controlled by an induction loop in the floor.

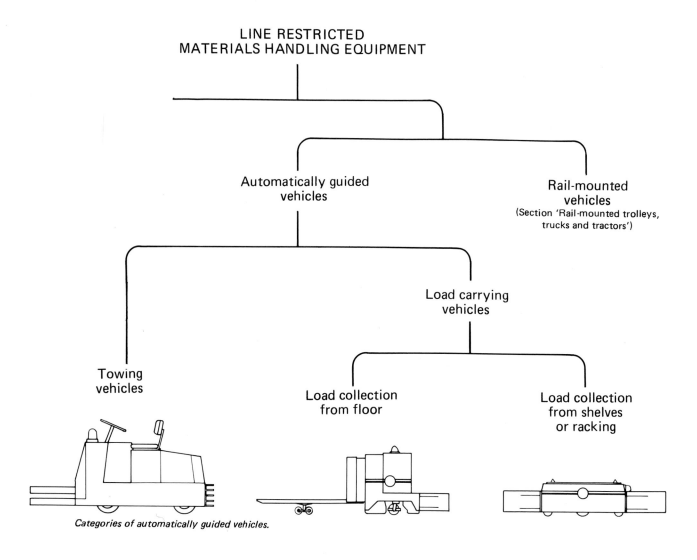

Categories of automatically guided vehicles.

Since the 1960s, towing AGVs have been in operation in factories, warehouses and terminals where their function is roughly that of towline systems (see section 'Rail-mounted trolleys, trucks and tractors'). Load-carrying AGVs were developed during the 1970s and will predominate during the 1980s.

Towing versions

Towing AGVs consist of a tractor and load-carrying trolleys coupled to form trains of 2—5 vehicles. The trains follow pre-set tracks (loops); stopping and vehicle uncoupling may be pre-programmed.

The load-carrying trolleys have platforms or roller tables conforming to the loading/unloading system used.

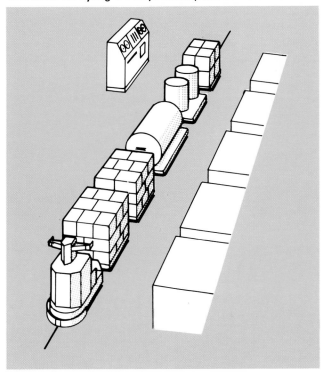

Towing AGV with platform trolleys for non-pallet loads.

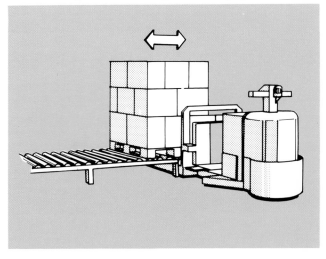

Platform trolley with roller table for automatic loading/unloading of pallet goods.

Tractor AGV speeds are 60—80 m/min and normal tractive power is up to a few tonnes. Minimum radius of bends is 2 metres, approximately, since trolley drawbars enable excellent tracking. The tractors can handle slight gradients and can also be used out-of-doors for transport between buildings. Equipment for automatic opening of doors etc., is available.

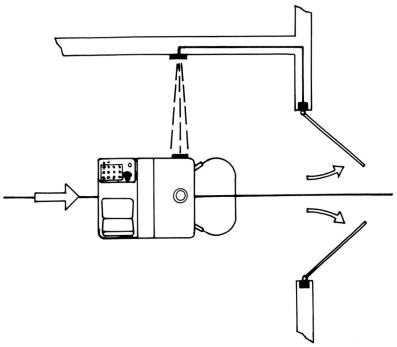

Doors can be opened automatically with the aid of photoelectric sensors activated by the tractor.

Examples of this type of installation include systems for supplying material to assembly lines, for collecting goods from various sections of spare parts departments and for moving finished products to automated storage installations (see section 'Case studies').

Load-carrying versions

This is the type of vehicle normally meant when reference is made to an AGV. Such vehicles may handle loads at floor level or from shelving.

Load-carrying AGV for collecting pallet loads from floor level.

Load-carrying AGV for collecting loads from shelves.

AGVs are particularly suitable for use in installations where greater flexibility is wanted, for instance, where layouts must be easily altered or floors left free for other traffic. Technical and economic developments in electronics over the last few years have also made AGVs a more attractive investment, reflected in an increasing number of suppliers. Some companies market both custom-made vehicles and conventional designs supple-

mented by automatic controls. Some designs are rather similar because of previous co-operation between manufacturers.

Floor collection versions are designed either as automatic forklifts or as automatic sideloaders. Modular design techniques permit some makes of base vehicle to be adapted to varying load capacities and geometry — handling everything from milk crates to pallets.

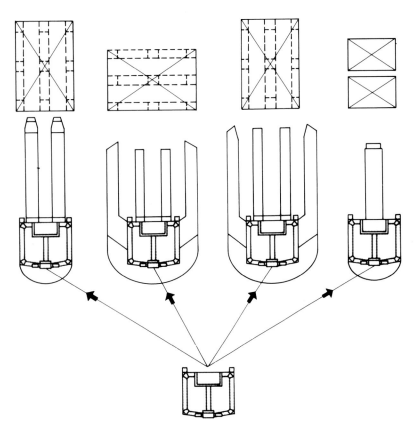

Modular system in which the base vehicle can be equipped with various fork attachments to fit different load sizes.

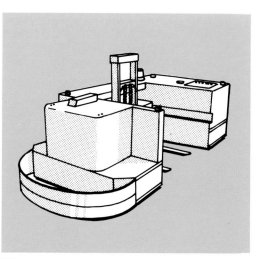

Sideloading AGV.

AGVs collecting goods from above floor level, for example, from shelves, conveyors, etc., are usually symmetrical front and rear. Loads are collected by lifting tables mounted on the vehicles. Sometimes, lifting tables are supplemented by powered roller-way or chain conveyor sections.

Loop controlled conventional forklift trucks, equipped with sensors at the fork ends, can handle high level stacking on their own.

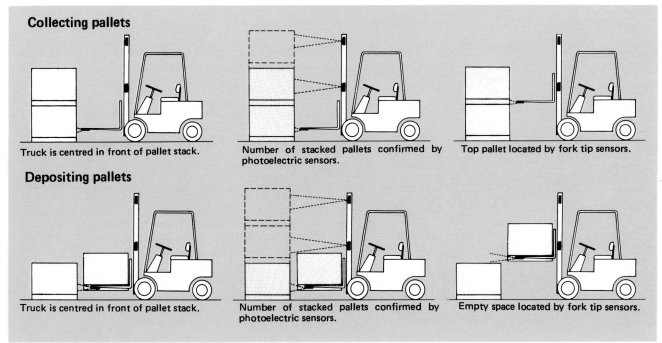

Collecting pallets

Truck is centred in front of pallet stack.

Number of stacked pallets confirmed by photoelectric sensors.

Top pallet located by fork tip sensors.

Depositing pallets

Truck is centred in front of pallet stack.

Number of stacked pallets confirmed by photoelectric sensors.

Empty space located by fork tip sensors.

Automatically controlled pallet stacking by AGV.

Typical specifications for an AGV may be as follows: capacity 1–1.5 tonnes, turning radius 0.6 metres, acceleration 0.5 m/sec² (from an emergency stop), and stopping accuracy when centring ±5 mm. Travel speed is 60 m/min., approximately, and time required for collecting or depositing goods is 5–50 sec. AGVs normally move as part of a one-way flow pattern but can be made to

reverse into stations to collect or deposit loads. Simple capacity requirement calculations can be performed by assessing the distance in metres which goods are to be moved in an hour, with added distances for empty running and for collection and setdown. More accurate calculations, especially for large systems, are performed using computer simulation.

Guidance systems

Towing as well as load-carrying AGVs may be equipped with guidance systems which follow floor markings (painted lines or glued-on reflective tape). Because such systems are vulnerable to mechanical wear and dirt, they are not used extensively.

Instead, magnetic reading heads at the front of

the vehicle are used in combination with induction loops in the floor. A minicomputer records jobs, determines routes, selects individual vehicles, monitors vehicle movement and determines flow priorities to avoid collision risks. Minicomputer commands are transmitted through substations to the AGVs which are usually equipped with microcomputers.

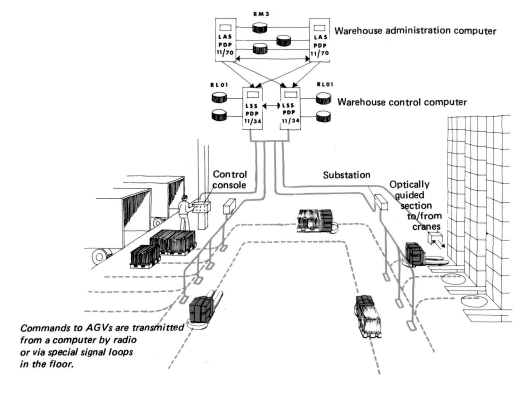

Warehouse administration computer

Warehouse control computer

Control console

Substation

Optically guided section to/from cranes

Commands to AGVs are transmitted from a computer by radio or via special signal loops in the floor.

The electric conductor in the floor carries a 3kHz A.C. signal and rests in a groove that is cut after the route has been decided. Joint filler or bitumen is used for refilling.

The magnetic field of the conductor is detected by one or several sensors connected directly to the AGV's drive/steer wheel. The steer wheel also carries the power unit. The vehicle will follow the loop until a switching point is reached.

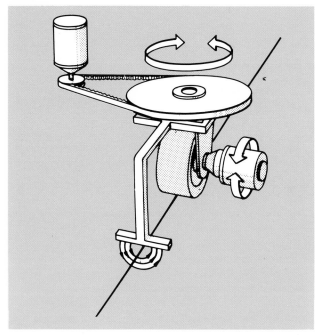

The magnetic field of the loop is sensed by a device connected to the drive/steer wheel.

Switching is based on one of several methods. The simplest involves cutting power to a main conductor section while energizing the side track loop. Another way is to use various fixed frequencies and to instruct vehicles to follow the main frequency until the selected side track frequency is found.

Different methods of **vehicle routeing** are also possible. One involves coding the information directly on to each AGV. The operator may manually enter the required destination on the vehicles, or equivalent data may be transmitted from a control system to each truck. Another alternative is to provide modulated frequencies from a separate communication loop in the floor. In each case, the trucks will transmit their coded destinations at each switching station, informing the control centre of their whereabouts and direction. When a vehicle destination coincides with the code for a given side track, the vehicle is switched to that side track. At the same time, a blocking signal, supplied to the main loop, forces the succeeding vehicle to wait until the way is clear.

Control can be exercised by a main minicomputer which keeps track of each truck position and transmits required control signals. Communication with an administration computer system may be via a VDU. Sub-programs may be entered, so that vehicles automatically proceed to a battery recharging station when no new job has been ordered or when batteries need recharging.

Some vehicle types equipped with microcomputers

are also able to leave the main loop. By measuring the direction of travel (the angle to the main loop) carefully and combining it with the distance travelled (recorded by wheel-actuated odometers) the track positions can be ascertained at any time. Vehicles, therefore, may leave the loop to allow overtaking, traverse floor sections with no guidance cable or enter cable-free floor areas to deposit or collect loads. In future it may also be possible to program such vehicles to deposit pallets on lorries.

Safety systems

Automatically guided vehicles are equipped with safety systems consisting of collision guard and winking lights. A collision guard is a sprung bumper that collapses under impact and triggers an emergency brake via a microswitch. Radar based detectors are now also available.

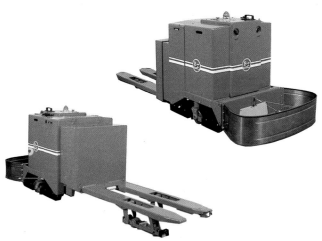

A radar detector at the fork tip warns of obstacles along the route and reduces vehicle speed. Sprung bumpers and winking lights are also included.

When necessary, vehicle controls can be overridden for conventional manual operation.

Application and development

Among examples are assembly AGVs with bodies specially designed for specific workpieces, e.g., car engines, car bodies or lorry cabs. Such vehicles are used, above all, as an aid for new factory layouts where assembly operations are carried out by self-contained teams (Volvo). The vehicles have speeds to suit individual stations as well as offering switching, accumulation, return in case of defects, simple transport of materials and tools, and central recording and monitoring of each unit.

AGVs have entered an intensive period of development because of their ability to offer flexible, labour saving transport systems in many branches of industry. Examples include mechanical engineering (Volvo, Saab), packaging (Tetra Pak), food processing (Rye King, Linköping Dairy) and large central warehousing installations (Oslo Goods Distribution Centre, Ikea) (see section 'Case studies'). Linked to computers, automatic storage cranes and production machinery, future AGVs will be employed in production systems with limited manning levels. Also, handling systems currently based on fixed conveyor lines (roller conveyors, transfer trolleys) will probably use AGVs.

Application examples

This vehicle is able to collect loads from fixed shelves and also to move in under food trolleys in a hospital.

AGV with roller table for pallet handling.

AGV designed to handle cars during assembly.

Low-lifting vehicle for handling loads between production plant and warehouse. Deposits and collects at floor or roller conveyor level.

AGV for handling heavy paper reels.

EXAMPLES OF AGV APPLICATION

	Towing version	Load-carrying version
Speed Capacity	60 m/min 2–4 trolleys each carrying 1,000 kg	60 m/min Normally 1 tonne (heavier loads possible in special cases)
Locating accuracy	–	±5 mm
Turning radius	2 m	0.6 m
Acceleration	–	0.5 m/sec²

Lifts and elevators

Lifts and elevators handle goods vertically (this excludes inclined conveyors) and may be divided into four main types:

Lifts (intermittent operation) have load-carrying cages or platforms which move up and down in vertical shafts or in frames with steel guideways.

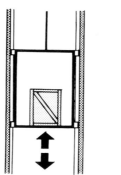

Lifting platforms (scissor lifts) are raised and lowered mechanically on the scissor legs principle.

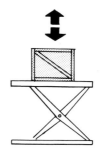

Paternosters (continuous operation) have several load-carrying units that move continuously or intermittently in an endless loop.

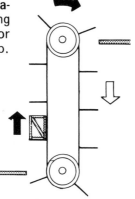

Platformless elevators have no individual load-carrying structures but allow random use. (Loads are moved by clamping or positive pressure).

Application

Lifts/elevators move goods between different levels, e.g., between conveyors at different heights or between storeys in warehouses and terminals. They can also bridge differences in levels which occur between new and old parts of an installation or between loading bay and ground level.

The same tasks may be performed by inclined (belt or slat) conveyors. Compared with them elevators require considerably less space but throughputs are normally reduced. For heavy and/or bulky loads lifts/elevators may be the only realistic solution.

Where sufficient space is available without an increase in costs, an inclined belt or slat conveyor is normally the cheapest and most efficient method of handling large flows of light or medium goods. Space and/or layouts are often decisive factors in choosing a lift/elevator.

Capacity and price vary greatly, i.e., from 10 kg for the smallest pneumatic elevator to 60 tons for the largest lift platforms, and from SEK 5,000 for the simplest lift platforms up to about one hundred times as much for a high capacity pallet elevator (1980).

192

Lifts

Lifts consist of a frame, platform or cage, hoist and enclosure.

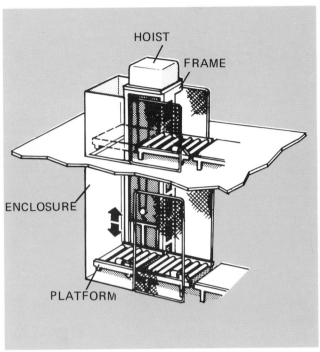

Principle of a lift.

There are many versions, from light duty models intended to lift light cardboard and wooden boxes short heights, to fully automatic equipment for pallet loads weighing up to several tonnes.

The frame may be a simple mast to one side of the platform or may have four corner posts which enclose the platform.

Loading/unloading may be manual or semi- or fully automatic. The platform may be equipped with a roller table, chain or belt conveyor.

Loading/unloading configurations may usually be chosen according to the alternatives shown in the figure.

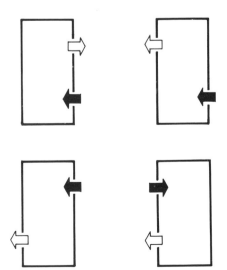

There are four loading/unloading configurations.

Loading/unloading sequences and vertical movements can be monitored by photoelectric cells and limit switches. Gates should be protected by wire mesh or automatically operated doors.

Power sources are either electric or hydraulic.

The illustrations show some alternative operating methods.

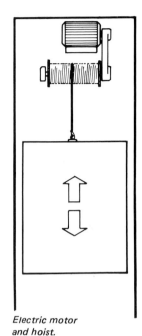

Electric motor and hoist.

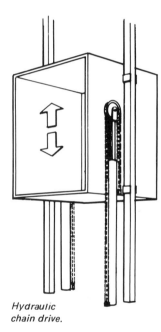

Hydraulic chain drive.

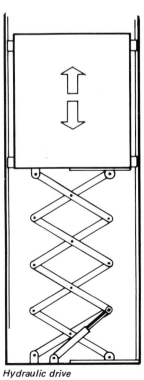

Hydraulic drive via scissor legs.

Alternative drive systems for lifts.

An electric hoist raises or lowers the lift cage via chains, steel cables or rack and pinion.

Chains are used primarily for slow speeds and loads up to 20 kg. Heavier duty, faster lifts employ an electric motor and cable hoist drum, an arrangement which needs space above to accommodate the hoist equipment and counterweights. Hydraulic drives usually require space for the hydraulic rams under the platform. On the other hand, motor and pump may then be positioned up to 30 metres away from the hoist — an advantage in terms of space utilization.

Rack and pinion drives may be advantageous for temporary installations (e.g., in building work, etc.) as drive unit and cage may then be of integral design.

For speeds of less than 1 m/sec, A.C. motors linked to worm or plain gear reduction units are used, although positional accuracy at stops can impose limitations. Acceptable positioning demands approach speeds not in excess of 0.2 m/sec. This requirement, together with the need for high hoist speeds, is met by using motors with 1:2 and 1:6 speed ratios.

For very great lift heights, more advanced equipment is necessary, for instance, a Ward Leonard unit of the type used in passenger lifts.

Paternosters

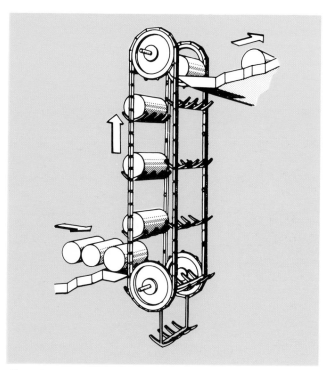

Operating principle of paternoster.

Unlike lifts, paternosters have a number of platforms or cages fixed to endless chains, etc. Operation is normally continuous. Loading/unloading (possibly automatic) occurs while the unit is running, resulting in greater throughput than with lifts plus a more even flow of goods.

Pallet elevators are a modified version of paternosters.

Light duty lift with wire mesh shaft. Used for handling bins, and wooden and cardboard boxes of up to 200 kg. Hoist speed is 15.5 m/min.

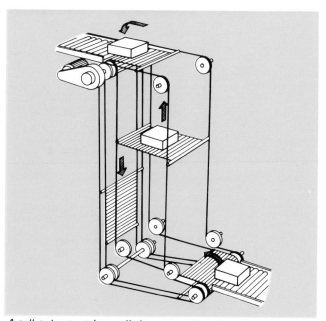

A pallet elevator takes up little space and handles pallets quickly.

Platforms consist of battens which form rigid horizontal surfaces in one direction of travel but behave like a roll-top during the return. This design saves space, is simple and offers high throughputs. Above all it is used for pallet transport between different levels. There are also special platforms for drums, bottles, paper reels, etc. Typical capacities are 8–16 pallets per minute, a figure which is reduced to 3–6 pallets per minute for pallets over 1,000 kg in weight.

Pallet elevator in a spare parts store.

Suspended tray conveyors have freely swinging platforms suspended from two parallel chains, permitting both vertical and horizontal transport.

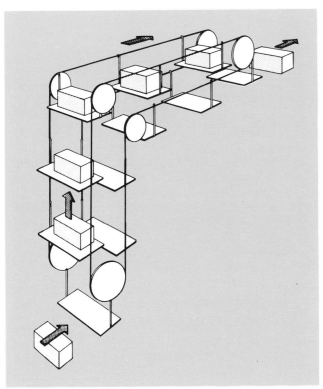

Suspended tray conveyors can offer both horizontal and vertical transport.

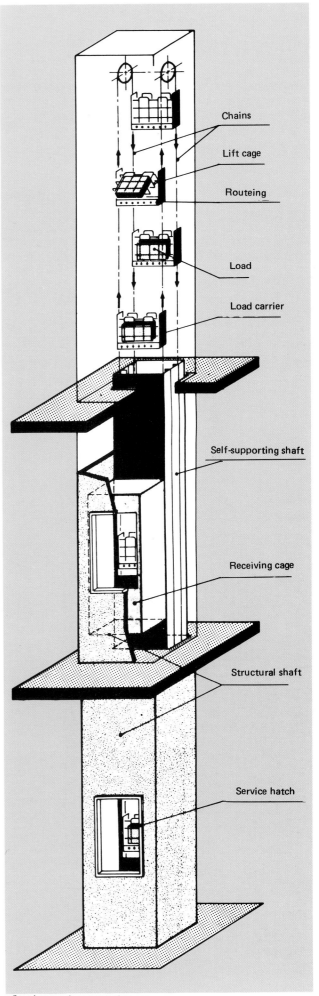

Continuous document elevator.

195

Platformless elevators

Mechanical platformless elevators are relatively new and were developed originally for vertical handling of mailbags and luggage at airports.

The load is held between two belts either by the belts' elasticity or by using an inflated air bag.

Both types are suitable for light general cargo. The former has the advantage in that it adapts its shape to the load and can therefore accommodate varying shapes and sizes. It is used only for lifting; lowering is via slideways or chutes.

In **pneumatic elevators** goods are moved through a tube by a difference in air pressure acting on their top and bottom surfaces. This method assumes uniform goods with plain sides which fit the tube.

Platformless elevator. An inflated air bag, containing the goods, is held between two flexible vertical belts.

Platformless elevator where goods are held mechanically.

Lifting platforms (scissor lifts)

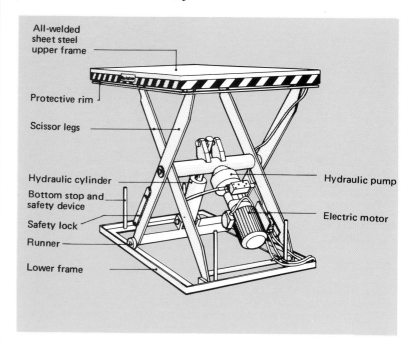

Lifting platforms are comparatively simple and versatile devices to bridge height differences from a few decimeters up to several metres.

They are slow compared with other elevators and are best suited to infrequent handling operations. They can be used instead of a bay for loading/unloading lorries, adjusting precisely to various deck heights. They are also used to move trucks from ground to bay level or between different levels which may be the result of extensions to old buildings.

Lately, lifting platforms equipped with traction motors and with wheels running on floor-level rails have been developed as order pickers in warehouses for lengthy goods (carpets, tubing, bars). They may then be an interesting alternative to more costly stacker cranes or to conventional warehouses designed for forklift or crane operations.

Lifting platforms are also extensively used in production, e.g., to feed or receive items from plant or to obtain proper working positions (as work tables, work platforms).

A lifting platform consists of a base frame and a top platform which is raised and lowered by one or several hydraulically operated scissor legs (see figure above).

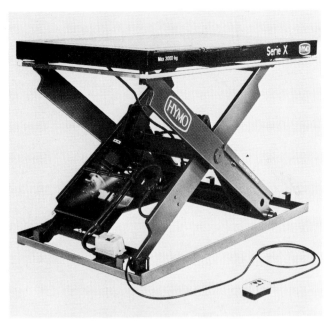

Platform with single scissor legs.

Platform with double (horizontally positional) scissor legs.

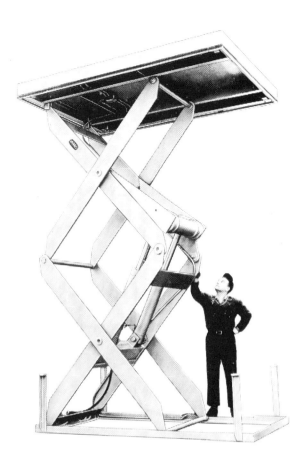

Platform with double (vertically positional) scissor legs.

Standard platforms are manufactured in three sizes. Single scissor-leg platforms are the most usual and are available in capacities of 500 kg–30 tonnes. Double scissor-leg (vertically positional) platforms are used for great lift heights (2–5 m). Double scissor-leg (horizontally positional) plat-

forms are used for lengthy goods and vehicles.

Lifting platforms may be supplemented by roller or wheel conveyors for simpler loading/unloading, e.g., in connection with feeding plant. They may also be mobile, with rubber wheels, rail-mounted wheels or air cushions.

Application examples

Pallet input to a high-level warehouse. Pallets are received and inspected on floor-mounted roller conveyors. A pallet elevator then lifts them to accumulating roller conveyors for transfer to high-level storage. Elevators were chosen to give the most economical utilization of available space.

When loading or unloading, a lifting platform may be used as a complete substitute for a loading bay. Lorry heights — varying from 0.8–1.6 m — are then immaterial but the work may take a little longer.

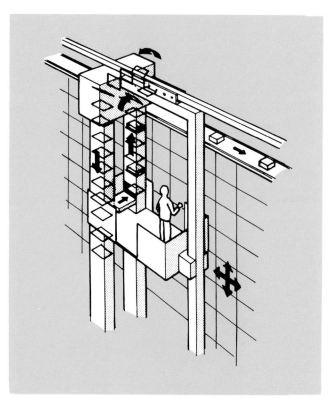

A manually operated crane in a high-level warehouse where the picker transfers goods from shelves to a paternoster which feeds a packing station via a belt conveyor.

A lifting platform may also permit an existing loading bay to accommodate various lorry heights without the use of ramps — an aid often necessary with conventional bays.

Steel bars in this warehouse are removed by hand but operations are facilitated with a double scissor-leg lifting platform equipped with traction motor and wheels running on rails (lift height is 2 m and travel speed 0.3 m/sec).

LIFT/ELEVATOR SPECIFICATIONS

Elevator type	Item weight kg	Number of items per min	Lift height m	Lifting speed m/min	Power source
Lift with cable hoist	50–2,000	1–3	Unlimited	12–24	El. motor
Pallet elevator	500–1,500	5–20	Unlimited	12–20	El. motor
Standard lifting platform, single scissor-legs	500–30,000	See lift speed	0.8–2.0	0.5–4	Electro-hydraulics
Lifting platform double scissor-legs (vertically positional)	750–8,000	See lift speed	1.5–5.0	2–4	Electro-hydraulics

Other line restricted equipment

This section covers various types of line restricted materials handling equipment which do not belong naturally under any previous heading.

Slideways and spiral chutes

The simplest practicable means to move goods from higher to lower levels is a slideway or a chute. The gradient of the line is balanced against the friction of the goods, resulting in a moderate speed which does not damage the goods. Such static devices normally accommodate light cardboard boxes, e.g., at postal and general cargo terminals or in order picking warehouses for multi-level handling.

Air freight terminal where spiral chutes are used to move goods from a higher (sorting) level to a lower (dispatch) level.

Straight slideways are used, for example, to complete a gravity roller conveyor line. Friction on the slideway brakes and stops the loads.

For especially lightweight goods spiral chutes may be made from steel wire rather than sheet steel, resulting in a very light and cheap design.

Document conveyors

Document conveyors are used to collect and distribute freight bills at goods terminals, to move orders and requisitions between workshops and warehouses or for internal distribution in offices, hospitals, etc. The fact that they are primarily document carriers, however does not exclude their being used to handle small items such as spare parts, samples, etc.

The three main types are pneumatic tubes, self-powered containers and narrow belt conveyors.

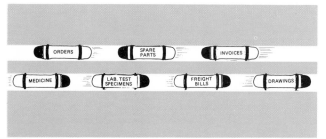

Transport by pneumatic tube.

Pneumatic tube station.

Pneumatic tubes are the most common. They are very fast (up to 600 m/min), reliable and simple to install. Documents or small components are housed in circular cartridges which are moved through a tube system by air power. Cartridges are available in diameters of 50–150 mm and in several lengths; the longest will accommodate A0 size documents.

Using these and other standard components such as pneumatic power sources, station switching devices and control centres, it is possible to construct systems of some complexity.

Direct document traffic is the simplest, and consists of a single tube, a station at each end and a power source.

Pneumatic tube cartridges.

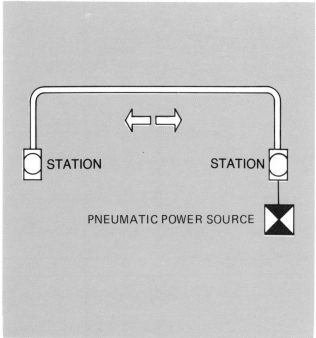

Direct system — the simplest possible pneumatic tube system.

Dispatch stations may be entirely manual or automatic. With the very simplest designs, the cartridges are put straight into the tube. More advanced designs include automatic dispatch by which cartridges are held until the system is ready to accept them.

The most complex systems may have up to about a hundred receiving stations offering complete intercommunication. There are also special systems for goods terminals and hospitals; several peripheral stations may then enjoy two-way communication with a central station though not with each other.

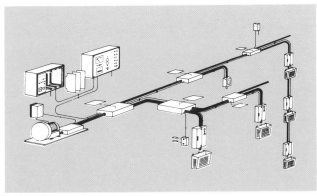

Complex pneumatic tube system.

Self-powered containers differ from pneumatic tube systems in that each container is self-powered and individually routed. The photograph shows a box equipped with its own electric traction motor and capable of finding its own way to the correct receiving station.

Self-powered container. Each container is individually coded to arrive at its correct destination.

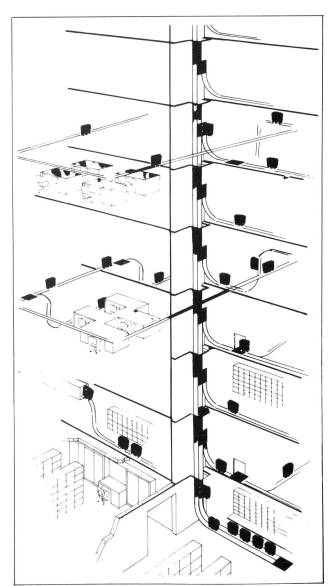

Self-powered container system.

Decentralized control and power supply ensure great flexibility and simplify later extension of tracks and stations. Systems with up to about a thousand receiving stations are possible. In comparison with pneumatic tubes, however, self-powered containers are considerably slower (30—40 m/min as opposed to 600 m/min).

Narrow belt conveyors are a simpler type of document conveyor, suitable for small systems, e.g., in banks, insurance offices, order departments and newspapers. The conveyor line consists of a U-channel above a moving belt. In the simplest cases documents may be put straight onto the belt to be conveyed along a single line. More complex branching systems may use coded cassettes on similar conveyor lines.

Narrow belt conveyor.

Air-operated conveyors

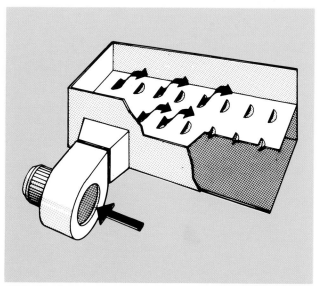

With the air-operated conveyor for very light loads, air jets from angled openings in the line supply the motive power. One application is as a conveyor for empty plastic beakers.

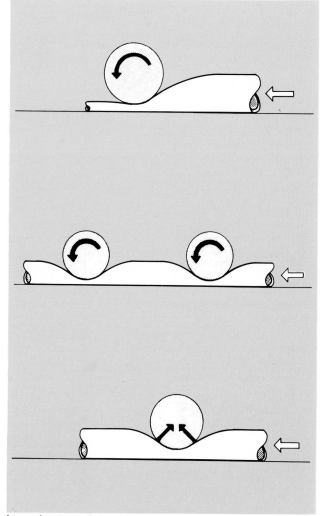

In another type of air-operated conveyor — still at the experimental stage (1980) — rollers are rotated via their contact with flexible hoses through which air is passed. Loads or load carriers can be placed directly on the rollers. The resulting conveyor can be compared to a roller flight, except that it cannot offer accumulation in the same way. Alternatively loads can be suspended from the rollers. In the resulting conveyor, the carrier is the only moving component.

Sorting installations

Materials have to be identified, labelled and sorted both in production and distribution. With manual handling this process is fairly simple as sorting staff can handle and inspect goods, decide on their destination and place them in the corresponding compartments. With mechanized handling, on the other hand, each unit must be labelled with an item or address code for machine reading into a control unit which in turn generates signals to mechanical switching devices (see section 'General information on line restricted materials handling equipment').

A range of devices are in use today, mainly in installations with large materials flow, e.g., in central depots for multiple retail stores (see section 'Case studies'), spare parts stores, mail-order firms, goods terminals and baggage handling installations at airports.

Choice of sorting system

A suitable sorting system (i.e., with reference to type and capacity) may be chosen on the basis of these factors:

- Physical characteristics of materials, such as size, weight, packaging, durability, etc. So, for instance, small unpackaged units require handling by belt conveyors and metal chutes while goods in cardboard boxes can also be conveyed on wheel or roller conveyors.
- Rate of flow, i.e., goods throughput per unit of time. Small rates of flow may be sorted by hand but where the rate exceeds 5—10 units per minute, mechanical/automatic handling may be better.
- Number of destinations. One or two operators can possibly manage a small number of destinations. Larger numbers may justify more advanced solutions because of space requirements and transport distances.
- Goods subdivisions required, e.g., each article by itself; a given number of units by themselves (e.g., in accumulation for automatic pallet loading); each order (destination) by itself.

Storage systems generally involve sorting of outgoing goods. These may be retrieved by one of two methods: order picking or batch picking. In the former case, goods are placed either directly on a designated load carrier (pick-and-place method) with no sorting equipment required, or transported to a sorting installation subsequently to be divided between the respective orders. In connection with the picking operation, goods have to be identified by some means, e.g., labels which are torn off the picking list and stuck on the goods.

In batch picking, goods are sorted in a similar manner except that several items of each article are picked.

Sorting in **production systems** is performed between different production stages whether some items are to be selected from a production accumulator for further processing or whether rejected components are to be removed after a sub-operation. In the former case, goods are sorted in the transport system; in the latter case sorting is usually a stage in production.

Design of sorting systems

A sorting system comprises:

Arrivals area
This is an area intended for accumulation and collection of goods waiting to proceed to sorting. In its simplest form, floor space is allocated for the reception of units prior to processing. Normally, however, it is desirable to save space and facilitate further handling by organizing the materials, for instance, using chutes.

Feed station
Here, each unit is separated for identification, destination and labelling.

Separation can be by hand, i.e., the operator takes one unit at a time, or mechanically, as when goods on a conveyor are transferred to another, faster conveyor line. To release one item at a time special arrangements can be incorporated on the basis of 'Stop and separation'.

Identification, in its simplest form, is the attachment of labels which can be read by humans. To facilitate reading and to reduce the risk of error, label information should be uniform and of sufficient size and contrast. Only short words, or groups of figures, should be used. Colour coded labels aid checking. Identification can also be automatic if reflecting tape, bar codes (UPC or EAN codes), etc., are employed. Codes are then read by photoelectric sensors or by a light pen.

Address allocation means that identified units are given destinations. An address may be the same as the identity of the item or a conveyor line.

Labelling means that the product — or that part of the conveyor which carries the product — is provided with the destination. The several methods are based on the type of control system used (cf. section 'General information on line restricted materials handling equipment').

In a control system based on routing information which accompanies the item, adjustable pins, printed squares, bar codes or magnetized spots ensure the appropriate action at switching points.

Where each item is monitored by a central system, pulses are generated to operate the required switches. Destination can be entered by hand through a keyboard, by voice labelling or automatically. Voice labelling is suitable where operators must use their hands to manipulate goods in order to see label texts.

Conveyor and switching elements
These include roller, wheel, chain, belt, slat and overhead rail conveyors (cf. respective sections). Switching points incorporate switches with lifting or tilting sections, deflectors, pushers, etc.

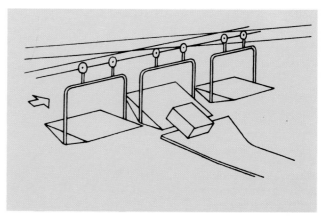

This overhead rail carries tilting platforms from which goods slide down into the appropriate collection area. Blocking time is zero as each item has its individual platform.

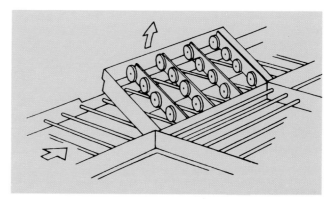

This design operates by gravity. The vertically reciprocating motion introduces the need to consider blocking times as previously described.

Examples of switching elements

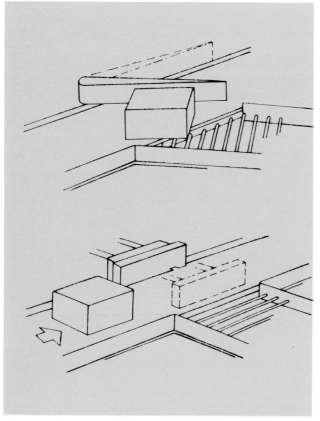

Here switching is accomplished by reciprocating arms or plates which force goods items off the line. In these cases the speed at which the devices return to their rest positions defines the blocking time. Blocking times are constant; thus increased conveyor speed does not increase throughput; the load zone lengths must be correspondingly increased.

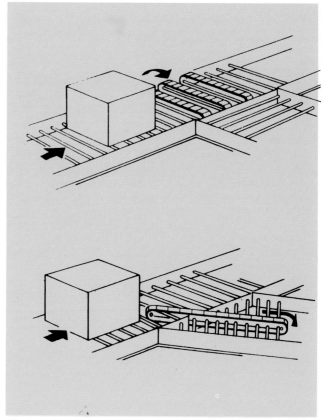

Designs showing variants of an identical principle. Wheels, belt or chains are raised between the conveyor rollers. Blocking again occurs, and blocking times depend on operating speeds of the switch mechanisms.

Capacities may be calculated using the indicated figures:

Parcel conveyor speed rates	
Type of conveyance	**Rate of speed m/sec**
Belt conveyor	0.1–1.0
Powered roller conveyor	0.1–0.55
Gravity roller conveyor	0.1–0.3
Accumulating powered roller conveyor	0.1–0.55
Slat conveyor	0.1–0.2
Overhead rail conveyor	0.1–0.5

Pallet conveyor speed rates	
Type of conveyance	**Rate of speed m/sec**
Powered roller conveyor	0.1–0.55
Chain conveyor	0.1–0.55
Overhead rail conveyor	0.1–0.3
Lifting transfer trolley	0.15 approx. Rest, braking and reaction time 7 sec approx.
Transfer trolley	0.3 approx. Rest, braking and reaction time 10 sec approx.
Elevator	0.1–0.4 Rest, braking and reaction time 16 sec approx.

Blocking times of parcel switching elements (Note the wide variations for different systems)	
Switching element	**Blocking time sec**
Lifting platform	3–15
Deflector	0–20
Pusher	0–20

Blocking times of pallet switching elements	
Switching element	**Blocking time sec**
Slewing unit	15–40
Lifting roller table	10–20
Transfer trolley	15–40

Collection areas

Sorted goods normally need an accumulation system upstream from packing and dispatch. Systems with limited flow use containers; larger systems use chutes or wheel/roller conveyors.

Examples of sorting installations

Roller and/or chain conveyors are used to sort **pallet loads**. Where there are many destinations, however, this solution becomes expensive (assembled switching units costing over SEK 30,000) and systems incorporating roller conveyors and transfer trolleys are then preferred (see 'Case studies').

Sorting installation for pallet loads. Accumulating powered roller conveyors form a return line loop for roll pallets. These have mechanically coded destinations and operate as input and output switching devices. The installation is fed from 7 lines, with discharge to 10 lines. Transport and sorting capacity is 450 pallets per hour.

This general cargo terminal uses a sorting installation for pallet loads which is based on a self-powered overhead monorail system designed for a hand pallet truck load of up to 1,500 kg. The drawbar between the carrier and the truck has a control box with three dials by which the destination is set. Speeds are 45 m/min maximum on the main line and 22.5 m/min on the sidings.

Parcels are sorted in a number of ways by equipment such as roller, wheel and belt conveyors (transportation) and deflectors, pushers and tilt devices (switching). Systems are constructed from standard components as well as from purpose-built elements.

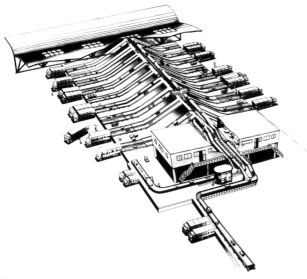

This sorting installation for cardboard boxes uses a steel belt conveyor with magnetic spot destination coding on the belt, immediately in front of each item.

Sorting by tilt-trays provides high throughputs (up to 10,000 items per hour), but sizes must be kept within tilt-tray dimensions. Standard dimensions are 500, 750 and 1,000 mm. Weights are normally up to 20 kg. Coding may be by pushbutton, optically (by pen reader) or acoustically (the operator imposes information via a microphone).

A sorting installation in a wholesale warehouse, handling bins containing order picked items. The bins carry optical bar codes which are read by a laser scanner. Destinations are transferred from the scanner to magnetic spots on the belt. Each spot actuates one of the 85 pushers switching bins to collection lines with rubber covered wheels. Sorting capacity is 6,000 bins an hour.

Sorting installation with tilting slats. The number of slats varies with the size of the goods items, providing greater flexibility and greater capacity (up to 5,000 items per hour). The installation can normally handle items weighing up to 50 kg.

Jib cranes and other fixed units

One way of handling heavy objects in confined spaces is by slewing cranes or load balancers. Slewing cranes have a long history, especially in engineering, while load balancers were developed in the 1970s. Slewing cranes are suitable for purely lifting operations within a larger area; load balancers are more suited to jobs involving combined lifting and assembly.

Wall-mounted slewing crane.

Slewing cranes

Slewing cranes consist of a slewing arm with travel beam and hoist. The hoist moves on the slewing arm and covers the area described by the arm's operational radius. These cranes are normally used indoors to serve various types of plant. In the past, they were often entirely hand operated but nowadays the hoist at least is powered. Common capacities are 50–1,000 kg. The hoists are basically similar to those described in the section 'Overhead travelling cranes'. In this context, chain hoists are the most common. Both electric and pneumatic operation is available.

Slewing cranes have the advantage over overhead travelling cranes in that maximum lifting capacity may be selected on the basis of the receiving capacity of associated processing machinery. By being fixed in one place a slewing crane is also at hand at all times and cannot be 'borrowed'.

Slewing cranes may be *column mounted* or *wall-mounted.*

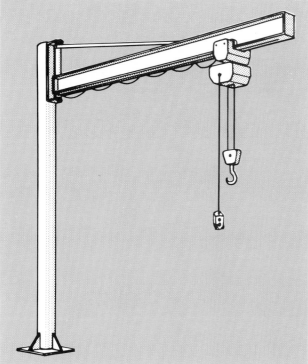

Column-mounted slewing cranes are fixed to the floor-base adjacent to the workstations they serve.

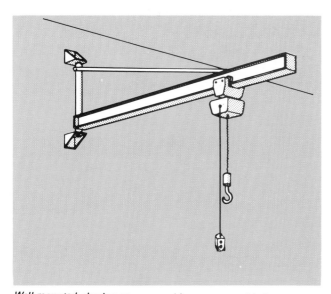

Wall-mounted slewing cranes provide operators with free working space.

With very heavy loads and great outreach, deflection may be troublesome and, therefore, slewing is sometimes powered. Column-mounted cranes accommodate load moments up to 5—10 tonne-force metres (maximum load X maximum outreach).

With column-mounted slewing cranes, a relatively small slew radius is often possible since the crane centre is adaptable to the machinery served. As a result, the crane can be comparatively low load-bearing and can be operated in conjunction with adjacent travelling cranes.

However, even moderate loads impose severe demands on the column-mounted crane's floor mounting and it is essential that baseplate dimensions are equal to the forces created and that crane positions are planned during the construction stage of the building.

With regard to wall-mounted cranes, the greater the distance between the wall mountings, the

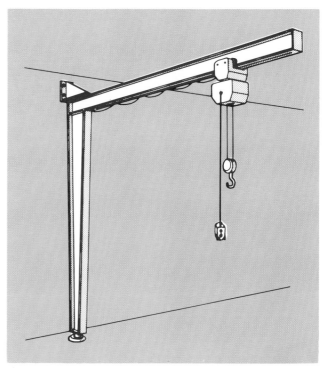

Strain and stress in crane structure and mountings can be reduced by combined wall- and floor-mounting.

smaller the reaction forces at the wall.

As a wall-mounted slewing crane is restricted to its working area, there is often a temptation to use comparatively large slew radii resulting in increased reaction forces at the wall. The wall mountings and the reaction forces they produce should therefore also be calculated and projected at an early stage to avoid later expensive reinforcement work.

Wall-mounted slewing cranes may incorporate a rail-mounted trolley to increase outreach.

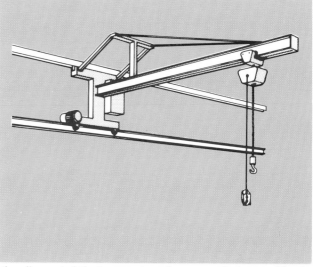

A wall-mounted slewing crane's work area may be increased by the addition of a trolley.

Load balancers

During the 1970s, several new aids to mechanized workpiece handling in production, assembly and storage were developed. They include units controlled by operators and programmable automated equipment. i.e., robots (see section 'Industrial robots').

The former group includes hoisting gear with attachments and continuous operator controlled motion. There are three variants:

- Power assisted manipulators with the operator seated at a control console. These units are usually designed to handle heavy objects in arduous environments such as foundries and forges.
- Special pick-and-place machines i.e. remotely controlled manipulators designed to pick goods in warehouses and stores.
- Load balancers with which the operator moves the workpiece by moving the hoist attachment. The operating sequence is as follows: the operator brings the attachment to the load, attaches it, switches on the load balancer control and moves the 'weightless' object to the position intended.

Load balancers consist of a movable jib with an attachment. The jib may be mounted on a fixed or mobile floor support, on a wall or at the end of a column-mounted slewing crane. Jibs are usually electrically or hydraulically operated, and the object handled remains at the same level when moved horizontally. Some makes are air operated.

Control is by manual movement of the supported load or by the operator controlling the jib's movements via a servomechanism.

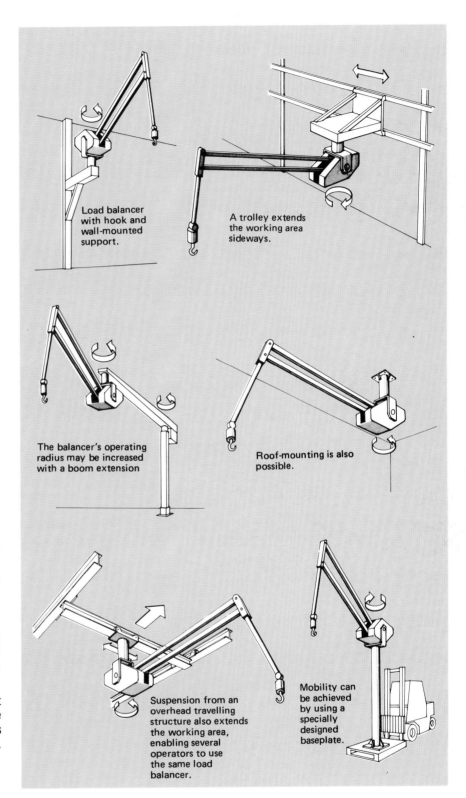

Load balancer with hook and wall-mounted support.

A trolley extends the working area sideways.

The balancer's operating radius may be increased with a boom extension

Roof-mounting is also possible.

Suspension from an overhead travelling structure also extends the working area, enabling several operators to use the same load balancer.

Mobility can be achieved by using a specially designed baseplate.

Attachments include slings, hooks, duplex tongs, electromagnets, suction devices and clamps. Automatic couplings simplify attachment change-overs.

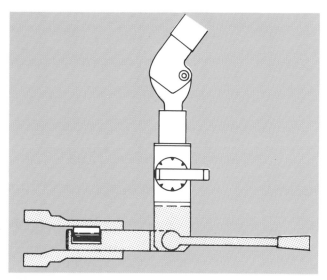

Attachment for engaging objects internally.

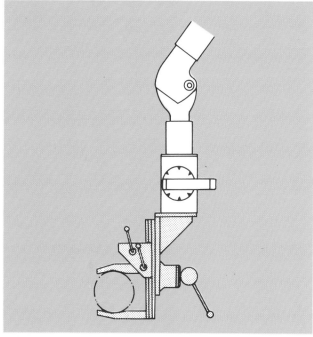

Attachment for external engagement of, for instance, a shaft.

The choice of load balancer depends on the weight and type of objects handled and the distances they have to be moved. Electrically and hydraulically operated units are capable of variable rates of speed regulated by a hand control. As a safeguard, the load balancers have a limit switch preventing the jib from moving towards areas of plant that may be vulnerable to damage. A brake or check valve stops loads from descending in cases of electricity or air supply failure. Some makes signal overloading visually or audibly.

Load balancers are likely to become a common handling aid since requirements for an improved working environment also mean that lifting weights permitted for manual operations will be reduced gradually. Existing installations indicate that, compared with manual lifting operations, capaci-

ties are reduced but improved ergonomics nevertheless create great advantages in terms of labour costs.

Manipulators

Remote control from a console fitted with a keyboard or servo-lever converts a load balancer into a manipulator. Such a handling aid is suitable where operators cannot approach the work area, for instance, because of high temperatures. One such manipulator is General Electric's 'Man-Mate'

Manipulator for handling a heated press forging.

with six degrees of freedom and 2,700 kg capacity. The machine can slew through 270°, its horizontal reach is 3.7 m and its vertical reach 6.3 m.

Manipulator for handling frame members in an automotive press line. It is of the same type as a hydraulic crane fitted to lorries for loading and unloading, with the addition of a pivoting hydraulic gripping attachment.

Pick-and-place machines are specialized manipulators comprising a rail-mounted vehicle with a driver's seat and space for a pallet, and an overhead frame with guides on both sides for a lifting device equipped with a gripping attachment (often suction cups). This attachment can be raised and lowered and rotated round its vertical axis. The driver picks goods from pallets positioned along the route and places them on the pallet on the machine (see pick-and-place machine in section 'Goods assembly and strapping equipment').

210

EXAMPLES OF SLEWING CRANE AND LOAD BALANCER SPECIFICATIONS

Slewing cranes						
Capacity kg	Hoisting speed m/min	Motor output kW	Slow hoisting kW	Hoist block height mm	Outreach m	Lift height m
125	8–13	0.5–0.8	0.1	500	up to 3	up to 3
500	4.5–8	0.5–0.8	0.1	500	up to 3	up to 3
2,500	4.2	4	0.9	700	–	–

Load balancers						
Make	Capacity kg	Power source	Hoist radius mm	Lift height mm	Hoist speed m/min	Approx. price (1980) SEK
A	110–280	Pneumatic	200–3,000	50–2,170	Manual	25,000 (160 kg)
B	75–225	Hydraulic	580–3,000	630–1,900	0–24	35–40,000
C	250	Hydraulic	380–2,580	285–1,800	0–28	45,000
D	30–400	Electric	450–3,250	300–3,300	0–20	25–50,000

Ranges refer to different models of some make.

Industrial robots

Industrial robots belong to the same category of mechanical handling equipment as load balancers but, unlike them, are capable of being programmed. Fixed and variable programs are possible.

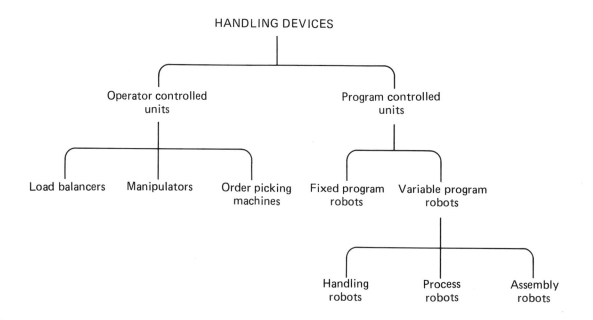

Industrial robots may be defined as an arm with a gripping attachment (both programmable) moving objects without human intervention between set points in space via a continuous path. Control movement may also follow set lines (contours). Thus there are two robot types:

Pick-and-place robots, i.e., units designed to move objects between two points in accordance with a set program. This group includes arms and gripping devices, and the equipment is specially designed to replace feed and discharge operators in press lines.

All-purpose industrial robots with programmable movement in up to 6—7 axes. Normally three axes are used to position the arm at a given spatial point, additional axes (when necessary) being used to locate the gripping device. Sometimes the work space can be extended by linear level, e.g., on a rail-mounted carriage. Experiments have also taken place with robots mounted on air cushion units and on automatically controlled vehicles.

The arm is of the extendable or parallelogram type, mounted on a central column with rotary mechanism and, where necessary, a raise/lower mechanism and tilting equipment.

Handling robot for feeding and discharging pressings.

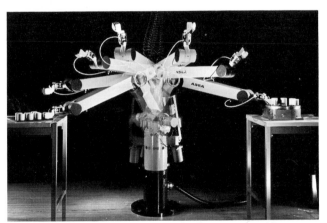

Arm and gripper movements are simultaneous, providing great flexibility.

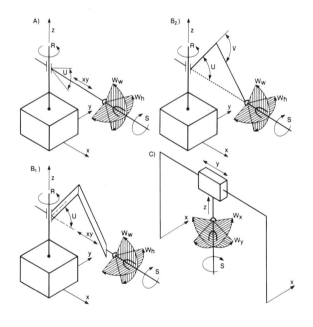

An extendable or linked arm mounted on a fixed or mobile base in combination with an articulated gripping device provides up to seven degrees of freedom.

All-purpose industrial robots may be classified according to application:

Handling robots to change the positions of objects and to serve plant. The former application may involve stacking vulnerable goods on pallets; the latter includes jobs such as handling workpieces in forges, feeding and discharging presses, pouring hot metal, unloading die castings and placing them in a trimming press, and serving machine tools either singly or in a group.

Process (or production) robots designed for activities such as handling spot welding guns, spray painting equipment and grinders.

Assembly robots. This type of robot is complex and still at the trial stage. Olivetti are testing their Sigma system for typewriter assembly, and General Motors fit carburettors using Unimation's new 3.5 kg Puma robot (Programmable Universal Manipulator for Assembly). The latter forms part of a manned assembly team.

Mechanical design

Industrial robots consist of four assemblies: base, arm, gripping device and control system. Design and combinations can be varied to produce variants with different characteristics and capacities.

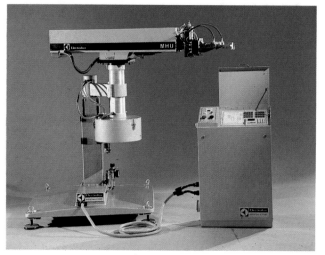

Industrial robot comprising base, arm (extendable type), gripping device and control system.

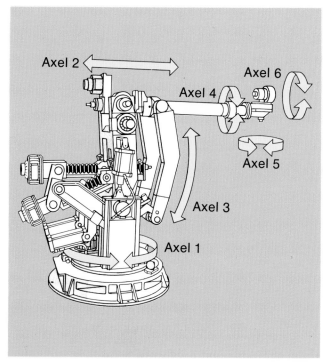

Industrial robot with linked arm.

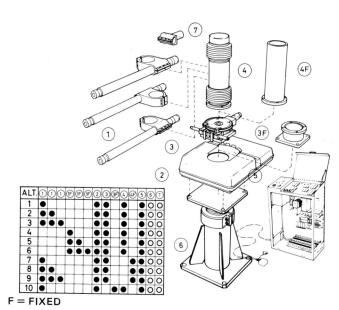

F = FIXED

Using modular units, industrial robots may be adapted to current operational requirements.

Motions are mechanically, electrically, hydraulically or pneumatically powered. Mechanical systems are usual for simple tasks such as feeding a press. Despite complex movement patterns electrical systems with servometers can provide simple solutions and are used by more and more manufacturers, even for weights up to 100 kg. Heavier objects may require hydraulic systems; pneumatics are used for simple movement patterns.

Mechanical drives are based on the use of recirculating ball screws, cams, eccentrics and chain transmissions. Electric drives use servo and stepping motors. Hydraulic and pneumatic systems have stroke and rotary cylinders, vane motors, etc.

Some industrial robots can be equipped with several arms and grippers, e.g., to serve a machining line by moving workpieces from machine to machine. Robot design principles facilitate modular construction of appropriate combinations.

Sometimes the gripper is mounted in a wrist-like device. This results in a wide variety of movement: in/out arm, arm raising/lowering, arm rotation, arm tilting, wrist rotation and wrist angling.

In practice, designs are limited to as few axes as is necessary because every additional axis brings with it mechanical, control and programming costs. Conversely, fewer axes mean more stable, more accurate, faster designs. Above all, the tilt capability has proved less useful on robots with built-in vertical arm movement.

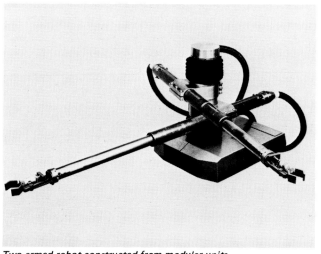

Two-armed robot constructed from modular units.

Half of the existing robots have fixed stop measuring systems providing ±0.1 to ±0.2 mm positioning accuracy; with electrical or hydraulic decoders, accuracy is ±1 to ±2 mm.

Most robots can handle weights up to 30 kg (there is no established capacity classification) although capacities do range up to 2,700 kg. In addition to acceptable handling capacity, robot specifications include the number of degrees of freedom, area of traverse for each degree, traversing speeds and positioning accuracies.

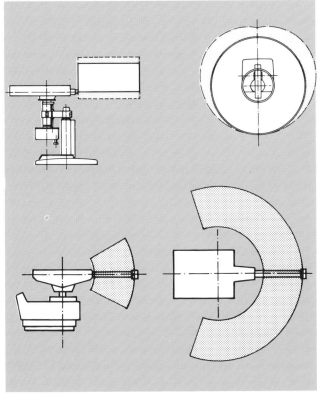

A robot's sphere of activity is determined by design and dimensions.

Control systems

Industrial robot control is based on *electromechanical, electronic* and *fluid pressure* systems. In the past electromechanical systems were predominant but electronic systems and mini/microcomputers are becoming more widely used.

Control systems have three main functions:
- Controlling and monitoring robot motions.
- Linking the robot and the machine it serves.
- Supervising the robot's sphere of operations to avoid accidents.

Control is usually arranged as a stepped program, the robot executing the program action by action. New actions are initiated by signal generators on the robot or peripheral equipment. The greatest difference between control systems is memory capacity, i.e., the number of programmable steps. Electromechanical systems are normally capable of 25–60 steps while electronic systems may allow several thousand steps (e.g., 127 programs at 127 steps = 16,129).

Monitoring functions include motor protection, pressure and temperature sensing. Communication with peripherals covers such functions as 'Workpiece ready for collection', 'Next machine vacant', etc.

The guard function may include an emergency stop actuated by photoelectric cells which signal intruders into the robot's working space.

A complete control system comprises five parts:
- A robot motion measuring system.
- A comparator.
- A memory store.
- A control unit.
- An activator.

The control unit informs the memory about the immediate program step. The memory provides command variable data for the various axes to be compared with actual values entered from the measuring system. Output signals from the comparator are transmitted to the activator which drives the robot until planned and actual values coincide.

The simplest form of control is *point-to-point* (PTP). The robot is controlled between programmed positions; intermediate stages or special traverses cannot be effected. This is still the most common type. Where the robot has to perform a continuous movement a *continuous path (contouring control)* system is used in which point-to-point control is combined with linear interpolation to produce a curved approximation. Such robots are used for qualified forms of materials handling such as arc welding, deburring, grinding, polishing and spray painting; for finishing, a magnetic tape controlled

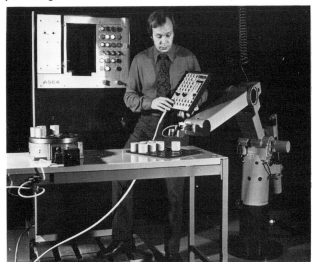

Programming an industrial robot with control console.

(playback programmed) system can also be used. Cams, permanent magnets and potentiometers are used as position indicators in electromechanical systems where stepping selectors and plugboards are used as memories. In electronic control systems measuring may be by a pulse counter (set by a visual scale) which is read by an optical (digital) system or a resolver or potentiometer (analog system). Punched tape, magnetic tape, drums, core storage or semiconductor storage may be used.

Gripping devices

The gripping device is an important (and sometimes costly) part of an industrial robot. It has to fit a given workpiece but should also be simply replaced to accommodate other objects, thus making the robot more flexible.

A gripping device should meet the following requirements:

- Accommodation of the shape, weight, size and surfaces of the object handled.
- Facilities for mounting on the robot arm, with drive and control system connections.
- Compatibility with pick-and-place stations (e.g., ability to operate in a confined space).
- Ability to grip despite dimensional and/or positional tolerance defects in an object.
- Low weight to minimise reduction in the robot's capacity, and to avoid increasing inertia moment and acceleration time.

The following types of gripping device are probably the most common:

- Internal and external clamps with two or more rigid or flexible fingers. The fingers are often replaceable and the clamping force adjustable.
- Vacuum grippers with suction cups. This type will operate only on fairly flat surfaces.
- Magnetic grippers for appropriate metallic objects. They are quieter than vacuum grippers and tolerate more demanding environments (e.g., workpieces covered in swarf and cutting coolant).

Grippers can be produced by taking casts of shaped objects. (Patent applied for.)

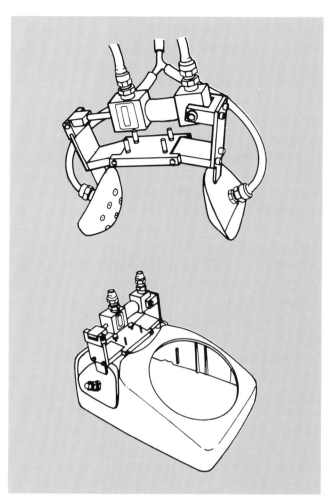

Gripping device based on casts taken of a telephone cover.

To grippers, may be added tools for cutting off casting stalks, separating components, etc. In the latter case, the grippers must be able to hold on to separated parts. Systems also exist where the robot itself changes grippers in roughly the same way as a multi-operation machine tool indexes tools.

Trials are also in progress in many places, with grippers able to sense the shape and size of an object and thus react accordingly. Sensing is by pressure indicator and photoelectric cell; TV cameras identify the appearance and position of the object.

Peripheral equipment

An industrial robot installation consists not only of a robot with gripping device and control system but also of peripheral equipment such as the plant to be served, magazines and conveyors. Designs are frequently determined by existing layouts and machinery. New premises, however, provide opportunities for extremely sophisticated solutions as a result of co-operation between designers and production engineers; robot-operated, unmanned vehicle-body welding lines now seen at several car manufacturers are but one example.

Gripper for cylindrical object.

Feeding workpieces to a robot may involve live or gravity operated conveyors. Because it is important that objects are properly located so that the robot may engage them as intended, magazine feed may be appropriate.

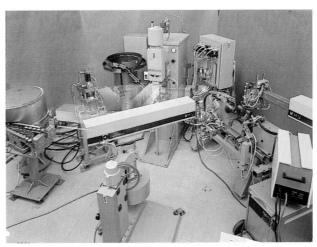

Industrial robots working in combination in a welding line.

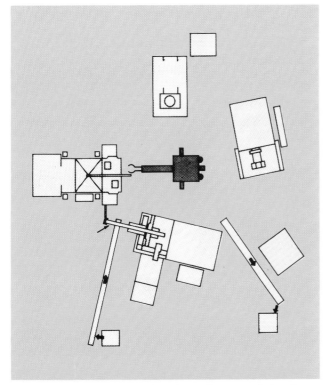

Robot operated installation. The robot collects a workpiece from a feed unit (see arrow) and moves it to a broaching machine, boring machine, lathe and balancing machine.

Application and development

Lately industrial robots have been much discussed in terms of technology, environment and social economy. Although originally conceived as labour saving automated devices for serving plant and for assembly, robots attract more and more attention as aids for work in awkward environments. Extensive development has been forecast and many quarters have expressed anxiety about the presumed effect of robots on employment in industry and trade. These misgivings have proved entirely without foundation, and robots installed to cope with repetitive, heavy jobs in severe working environments are seen as positive gains by all parties involved.

At present (1980) approximately one thousand industrial robots are thought to be in use in Sweden (including pick-and-place units). One of the largest users is Electrolux, a company which initially developed a model for in-house operation but later marketed it to other companies. Electrolux's plant at Motala has installed around seventy

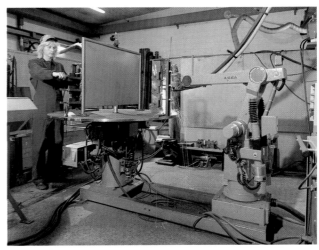

Arc welding robot.

robots to serve, e.g., injection moulding machines in two shifts.

Volvo uses industrial robots for, among other things, welding car bodies. For example, a six-robot installation performs 340 spot welding operations in 72 seconds.

Many West German, French, U.S. and Japanese robot manufaturers' products are now in use in Sweden.

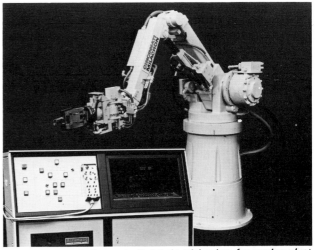

Six axis numerically controlled industrial robot for serving plant and for welding.

Today (1980) the price of an industrial robot, including control system, is between/SEK 100,000 and 450,000, depending on size, gripper device, control system, etc. To this must be added costs of altering and adapting plant and feed systems, increasing the cost of a robot operated work station to SEK 500,000–600,000.

Future generations of industrial robots are likely to be equipped with visual and touch sensors to identify different objects. Precise positioning of objects will no longer be required and robots may select different handling programs according to workpiece shape. These features will be especially valuable in assembly.

Intelligent, computer controlled industrial robots will more and more work in groups together with other robots and plant, forming large production systems within unmanned factories.

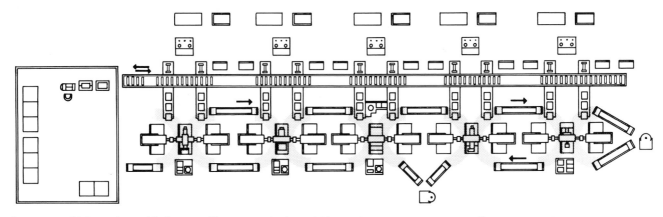

Japanese machining and assembly line served by ten portal robots which transfer workpieces between different machine feed magazines.

As already stated, a large number of robot makes are now available.
Specifications for two are given below:

EXAMPLES OF INDUSTRIAL ROBOT SPECIFICATIONS

Make	A		B	
Capacity	6 kg	60 kg	5 kg	15 kg
Power	Electric	Electric	Pneumatic	Pneumatic
Horizontal reach	1,200 mm	2,200 mm	500 mm	1,100 mm
Vertical reach	1,600 mm	2,200 mm	150 mm	500 mm
Maximum angle of rotation	340°	330°	270°	$\pm n \times 360°$
Arm rotation speed	95°/sec	90°/sec	180°/sec	90°/sec
Arm in/out speed	750 mm/sec	1,000 mm/sec	1,000 mm/sec	1,000 mm/sec
Arm up/down speed	1,100 mm/sec	1,500 mm/sec	500 mm/sec	300 mm/sec
Positioning accuracy	±0.2 mm	0.4 mm	±0.1 mm	0.5 mm

Note that most suppliers offer different models for different capacity requirements. The above figures exemplify approximate applications.

Load carriers (pallets, stillages, etc.)

Load carriers are normally structures holding together one of several products, facilitating handling and storage by standardized means. They are made in different sizes and from different materials determined by the size, weight, geometry, environmental requirements, etc., of the goods handled. This, in combination with various strapping methods, results in load carriers ranging from simple wooden supports to expensive thermally insulated boxes.

Load carriers are designed to be compatible with existing storage and transport systems (and vice versa). The dimensions should be such as to accommodate large and small units which fit into a modular, standardized system. In this way **unit load** systems are obtained.

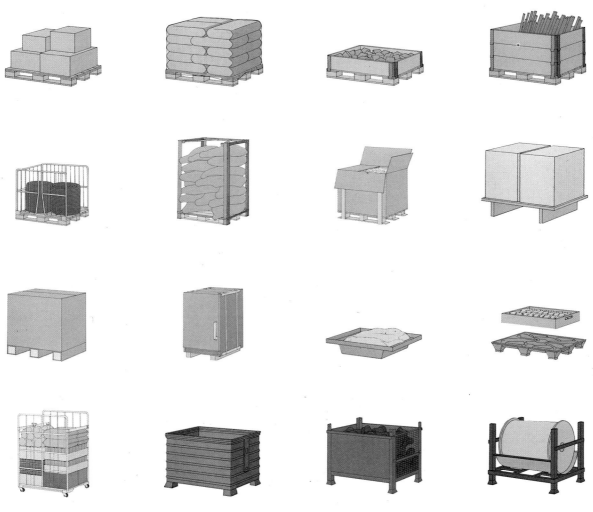

Examples of unit loads.

Product dimensions should obviously, where possible, be adapted to the envisaged unit loads. A lump of sugar is a good example: its dimensions are contained in consumer packages, wholesale boxes and pallets. In other cases compromises must be accepted, as when handling eggs and egg containers.

Unit loads should also be composed so that they move intact, through multiple transport stages, for as long as possible.

A good rule to follow is:
Load unit = Transport unit = Storage unit.

Load carrier classification

One way to classify load carriers is in order of increased size and complexity:

- Loose supports (pads) which are strapped to goods to provide storage protection and access for handling (e.g., by forklifts).
- Simple boards of wood, metal or plastic which hold goods together and facilitate roller conveyor handling.
- Bins, boxes and baskets for manual handling, constructed from wood, plastic, sheet metal or wire mesh, which hold goods together and protect them.
- Bins and boxes for mechanized handling. Basically similar to the above, they may have a smooth underside, slots to facilitate lifting and be capable of carrying a routing code.
- Disposable or re-usable pallets designed for fork handling and transport along mechanized lines. Various designs are possible and materials used include cardboard, wood, plastic, foam, aluminium and sheet steel. Folding frames (converters) to build pallet cages are available.
- Roll pallets, normally wooden pallets with wheels and wire mesh sides and gates.
- Large open platforms for crane or truck handling at timber terminals and warehouse installations. The platforms may be transported by rail, road or sea and deposited on supports (demountable bodies) or on the ground (container platforms or demountables).
- Demountables with box bodies (one-piece or with tarpaulin covers), often equipped with leg supports and channels to facilitate handling by vehicles equipped with simple chassis-mounted lifting equipment.
- ISO containers. Equipped with corner fittings and often made of steel. Dimensions and capacities comply with international standards (see below). Acceptable to all transportation systems; can be handled by crane or truck (normally using spreaders) and stacked several containers high.
- Special purpose containers such as semi-rounded units designed to fit aircraft fuselages.

Bins

Bins, boxes and baskets are available in many versions. Common to most are the modular base measurements of 400 × 600 mm in conformity with Euro pallets (see 'Pallets and pallet accessories'). Sometimes, the shape is slightly tapered and incorporates feet enabling the units to be stacked on top of each other when full and to rest, when empty, in each other to save space. For storage purposes there are also long, narrow boxes, e.g., 50 mm wide and 300 mm long. Height is 50—100 mm and adjustable partitions are often fitted. Grooves for handling label holders are common.

Plywood, cardboard, sheet metal and plastic are usual box materials. Mechanized installations may require special solutions, e.g., completely smooth

Plastic bins.

Plywood boxes.

undersides for unimpeded use on roller conveyors or reinforced leading edges where the boxes engage crane attachments in automated warehouses (such as automated miniwarehouses which contain archives, libraries, banks, vaults, spare part stores, etc.). Mechanized systems may also have equipment for stacking and separating bins and boxes automatically. Among places where such equipment is used are dairies.

Plastic bins on shelving.

Pallets and pallet accessories

Pallets are, without doubt, the most common carrier of unit loads. They were developed concurrently with forklift trucks in the 1940s, and now form the basis of all rational materials handling.

Pallets come in many different designs and may be made from wood, cardboard, plastic, aluminium, steel, combinations of materials, etc.

The illustrations below show different designs:

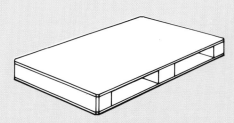

Two-way entry pallet — handled from two sides only (common in the USA).

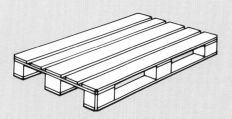

Four-way entry pallet (most common type in Sweden).

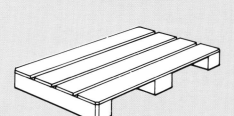

Single decked flat pallet — one deck and bearers.

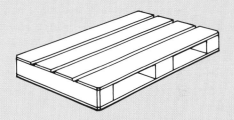

Double decked non-reversible pallet.

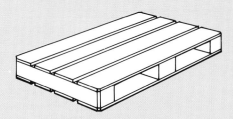

Reversible pallet — top or bottom deck can carry the load.

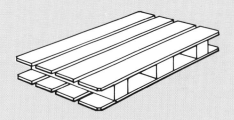

Wing pallet — in which one or two decks project beyond the spacers. Used for sling handling, e.g., when loading/unloading ships.

A distinction is made between re-usable and disposable pallets. Re-usable pallets include the standardized Euro pallet (essentially a European exchange pallet), predominant in Scandinavia and in countries such as Germany, Belgium and Switzerland. An estimated 11 million are used in Sweden (see also TFK Report 80:7, 'Pallet management').

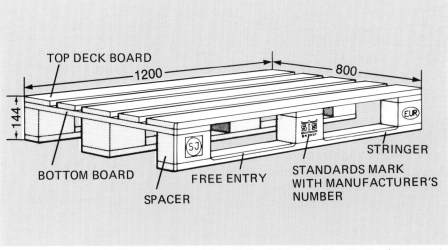

European exchange pallet (Euro pallet) complying with Swedish Standard Specification (SS 842007).

221

The Euro pallet is standardized in Europe with regard to size, material and construction, including nailing. On the basis of this standard, European countries have agreed to exchange Euro pallets between their respective railway administrations and between various domestic transport users in countries such as Germany, Norway and Sweden.

In addition to the 800 × 1,200 mm standard Euro pallet, there are several other standard sizes:

- 400 × 500 mm builder's pallets (SS 841011).
- 600 × 800 mm half-pallets (SS 842004).
- 1,200 × 1,600 mm marine pallets (SS 842006).

Euro pallets fit ISO containers badly (see figure below) and a 20 ft container can accommodate only eleven.

This has resulted in long discussions at ISO, the International Organization for Standardization, about proposed international standardization of pallet dimensions. As a result a compromise proposal, based on the 1,000 × 1,200 mm pallet size, envisages satisfactory container filling, using an irregular loading pattern. This proposal has still (1980) been unacceptable to countries such as the USA and Japan where the 1100 mm dimension is preferred, i.e., 1100 × 825 mm, 1000 × 1100 mm, etc.

In spite of this there is much to indicate that the 1000 × 1200 mm pallet will grow in importance, at least in Europe. It offers opportunities for producing a stronger pallet which would enable vehicles to be filled more completely. At the same time it conforms to the packing modules (400 × 600 mm) developed in the European food industry.

Examples of special pallets		
Pallet type	**Size mm**	**Remarks**
Swedish brewer's pallet	1080×1080	Adapted to beer crates.
Volvo pallet	1225×820	Adapted to open demountables.
Supra pallet	1050×1350	For fertilizers packed in bags.
Scania pallet	800×1200	Equivalent to Euro pallet but with one-piece decks.
PLM pallet	1200×1350	Adapted to empty cans of low weight by volume.

Plastic pallet for food handling.

More demanding environments (food processing, hospitals, etc.) may require plastic or aluminium pallets (costing SEK 200—300 each). Such pallets will stand cleaning with superheated water or steam.

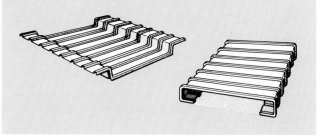

Steel pallets.

There are also pallets constructed either from simple, profiled sheet steel (easy to stack when empty) or welded box fabrications.

In addition there are disposable pallets in cheaper materials (cardboard, low grade wood, etc.) which are convenient where retrieval difficulties arise, e.g., supplying materials to building sites or exporting to countries outside the 'pallet pool'.

Pallet type		Number of pallets	Area utilization
800 × 1200 Euro pallet		11	77.4%
1100 × 825 Container adapted pallet		14	92.7%
1000 × 1200 Proposed international standard		10	87.6%

5 867 · 2 330

Floor area utilization when loading different standard pallets in an ISO container.

In addition to the Euro pallet, there is a range of different special purpose disposable pallets. Some large organizations have created their own internal standards. The reason for such sizes may rest with durability and quality requirements or the desire for a closed-circuit, controlled system of owner-marked pallets which cannot be mistaken for, or exchanged with, Euro pallets. The table shows some examples:

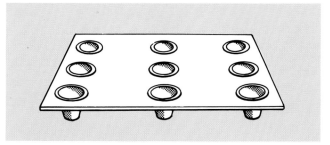

Example of one-way pallet.

Pallet collars add side walls to standard pallets. They are usually 200 mm high; hinges enable them to be folded for return or storage.

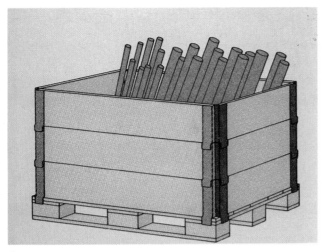

Pallet fitted with collars.

Pallet converters consist of metal frames and gates which can be clamped to pallets, enabling picking from block stacked pallets.

Pallet converters may be mounted on a pallet's long sides.

Gates mounted on the short sides create a complete box pallet.

Finally, it may be enough to clamp simple corner posts, stabilized by some kind of cross-bracing at the top. Pallets can then be block stacked while containing overhanging loads (such as animal carcases in cold-stores).

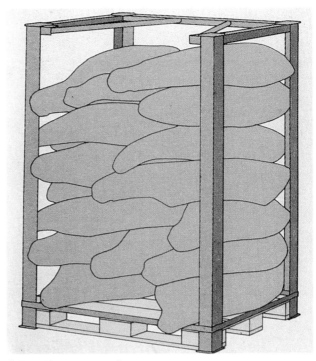

Pallet with corner posts for horizontally stored carcases.

Roll pallets

This device may be regarded as a wheeled variant of the timber pallet/converter combination and has been developed as an aid to distributing everyday household goods and groceries. Used with a lorry tail-lift, a roll pallet provides a logical delivery method to retail outlets where the standards of goods reception may vary.

Roll pallets are available in various sizes, mostly as modules of the Euro pallet's basic 800 × 1200 mm. A half-pallet size base (inner dimensions ~600 × 800 mm, outer dimensions 700 × 800 mm) is the most common and is used by all food/goods wholesalers.

Roll pallet for distributing mixed loads of everyday goods from wholesaler. Usual external size 700 × 800 mm.

A special stainless steel pallet in quarter-pallet size (600 × 400 mm) has been developed for milk and other dairy products; it can accommodate 180 litres.

Special roll pallets for suspended carcases have also been developed.

The combination of roll pallet and tail-lift has become a must in the distribution of food/goods in cities.

223

Containers and demountable bodies

This equipment group includes:

- Demountables (Swedish Standard SMS 3038, 3039, 3043).
- Containers (SMS 842100–842108; ISO standard design).
- Container platforms (SMS 842010 draft standard).
- Semi-live lorry bodies/containers (SMS 3021).

Standards specify:

- General dimensions.
- Dimensions associated with handling and securing on vehicles.
- Maximum permitted gross weights.
- Handling/securing requirements and methods.
- Strength requirements with test methods.

Demountable bodies

A demountable body consists of a frame with legs, possibly also with some kind of superstructure. It may be a platform with flaps (with or without tarpaulin cover) or a box body. Demountables are designed specifically to facilitate simple on/offloading where vehicles have no auxiliary aids. They frequently have guides underneath to ease reversing when being loaded onto vehicles.

Regulations for securing loads to load carriers and load carriers to vehicles have been published by Swedish Railways (SJF 620:1 and 638:1) and the National Swedish Shipping and Navigation Administration (Ordinance 1973:A 14).

Containers and demountables permit unit loads to be handled through storage/transportation sequences by road, rail, sea and air. Such systems are becoming more and more common and essential for several reasons:

- Transhipment times (terminal times) are shorter; expensive vehicles need not remain unused during loading and unloading.
- Vehicles can be used more extensively: for instance, in summer a lorry can be used for ice cream distribution and in winter to sand icy roads.
- Transport routes for different products can be planned using the same vehicles.

This has led to rapid development of containers and demountables since the 1970s.

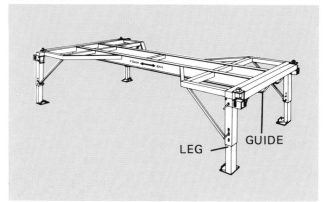

Demountable body for container transport.

In accordance with SMS 3039 demountable bodies are recommended in three sizes and two weight classes as below:

Type	Length m		Width m		Height, excl. legs, of platform with cover m		Max. gross weight kg	Tare weight approx. kg		Max. capacity approx. kg		Load volume for platform with cover approx. m³
	Ext.	Int.	Ext.	Int.	Ext.	Int.		Platform	Platform with cover, fork pockets, support brackets	Platform	Platform with cover, fork pockets, support brackets	
100	6.06	5.0	2.50*	2.43	2.50	2.20	16,000	1,900	2,250	14,100	13,750	31.5
163	6.00	6.45			2.50	2.20		2,100	2,450	13,900	13,550	34.5
16A	7.15	7.00			2.50	2.20		2,300	2,700	13,700	13,200	37.4
16A	7.15	7.00			2.80	2.50		2,400	2,800	13,600	13,200	42.5
10C	6.06	5.90	2.50*	2.43	2.50	2.20	10,000	1,800	2,150	8,280	7,650	31.5
10B	6.60	6.45			2.50	2.20		2,000	2,350	8,000	7,650	34.5
10A	7.15	7.00			2.50	2.20		2,200	2,600	7,800	7,600	37.4
10A	7.15	7.00			2.80	2.50		2,300	2,700	7,700	7,600	42.5

Remarks: — Normal external height 2.5—2.6 m
— Normal flap height 0.6—1 m.
— Where a 1 tonne forklift is used for loading, floors should be designed for not less than 3,500 kg axle weight.
Note: From Nov., 1980, max. permitted vehicle width in Sweden is 2.60 m. New demountables conforming to this width may therefore be expected.

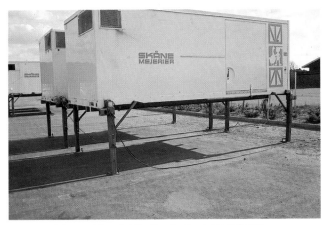

Demountables used for distribution in the dairy industry.

Demountable with flaps and cover.

Containers

In this context, 'container' implies the international standard recommended by ISO and should be named 'ISO container'. This standard stipulates a range of length dimensions (5, 10, 20, 30 and 40 ft) but one common width of 8 ft (2.44 m), and special corner fittings for handling and securing.

The ISO container has gained extensive popularity, primarily in shipping. The most common size is the 20 ft unit of which there are about 2 million (1981).

A container consists of floor, walls and roof. It is normally opened at the end but some units designed for special products permit entry from the side or top. Containers may also have insulated walls and refrigeration units.

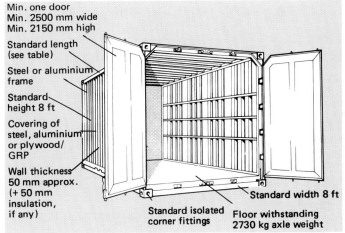

Min. one door
Min. 2500 mm wide
Min. 2150 mm high

Standard length (see table)

Steel or aluminium frame

Standard height 8 ft

Covering of steel, aluminium or plywood/ GRP

Wall thickness 50 mm approx. (+ 50 mm insulation, if any)

Standard isolated corner fittings

Standard width 8 ft

Floor withstanding 2730 kg axle weight

ISO container with end doors.

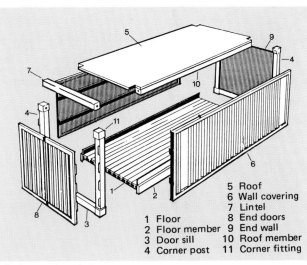

5 Roof
6 Wall covering
7 Lintel
8 End doors
9 End wall
10 Roof member
11 Corner fitting

1 Floor
2 Floor member
3 Door sill
4 Corner post

Basic container design.

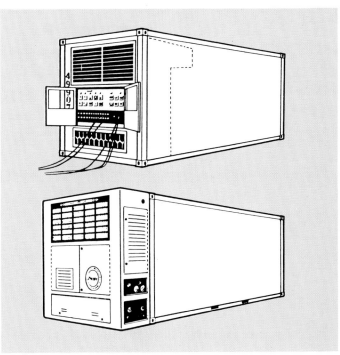

Refrigerated container.

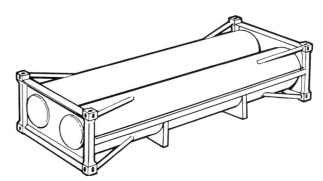

4 ft. high container. Other dimensions in accordance with ISO standard. One stacked on top of another equals normal container height. Such half-height containers are used for goods of high weight by volume.

Containers must be stowed properly for sea transport. In heavy seas, a ship may roll as much as 45°. If goods are not well stowed and secured, they may shift and suffer damage.

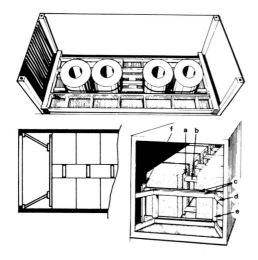

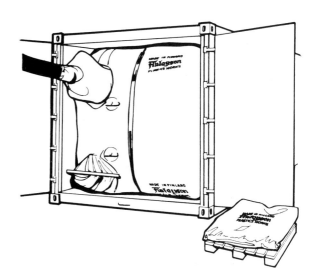

When stowing goods in a container, many different aids and methods are available. Factors such as condensation, odours, temperature sensitivity, etc., must be allowed for. Advice and instructions for container stowing is given in TFK Notice No. 64, 'Stowing goods in containers on platforms', and TFK Report 1978:3, 'Stowing heavy goods in containers and on platforms'.

Rubber envelope in standard container for one-way transport of, say, powdered goods. On the return journey the envelope is folded and the container may be used conventionally.

Containers are available in four lengths: 10 ft, 20 ft, 30 ft and 40 ft. In Sweden they are mostly 20 ft and 40 ft. The dimensions can be seen in the table.

ISO container specifications										
Type	Length m		Width m		Height m		Max. gross weight kg	Tare weight approx. kg	Max. capacity approx. kg	Load volume m^3
	Ext.	Int.	Ext.	Int.	Ext.	Int.				
1 A Closed					2.44	2.20		3,700	26,780	61.5
1 AA Closed	12.19	12.00	2.44	2.33	2.59	2.35	30,480	3,800	26,680	65.7
1 AA Open top					2.59	2.32		4,200	26,200	30.1
1 C Closed					2.44	2.20		2,100	18,120	30.1
1 CC Closed	6.06	5.87	2.44	2.33	2.59	2.35	20,320	2,300	18.020	32.1
1 CC Open top					2.59	2.32		2,500	17,820	31.7

Remarks: — Tare weights indicated refer to steel structures.
— Tare weights may vary between different makes.
— When tested, floors should withstand a max. axle weight of 5460 kg.

Container platforms

A container platform is basically an ISO container base with or without end walls. The dimensions are:

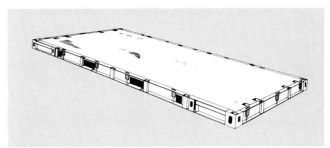

Container platform.

Container platform specifications						
Type	Length m	Width m	Height m	Max. gross weight kg	Tare weight kg	Max. capacity kg
Base only	6.06	2.44	0.26	20,300	1,700	18.600
With folding end walls	6.06	2.44	2.59	20,300	2,800	17,500
With fixed end walls	6.06	2.44	2.59	20,300	2,800	17.500

Semi-live lorry body

Semi-live lorry bodies have a load carrying frame designed to be pulled onto a vehicle from the ground. Superstructures (refuse containers, tanks, etc.) are inclined up to 25° which is acceptable if normal securing regulations are followed. Units can also be fitted with legs.

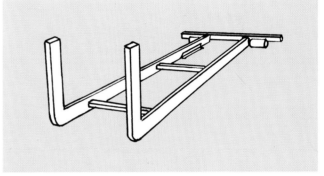

Semi-live lorry body frame.

Handling a semi-live lorry body by hydraulics.

Other variants

The range of variants includes:

- **SJ railway containers** measuring 7000 × 2420 mm internally. They are designed primarily for Swedish domestic use and — compared with 20 ft containers — provide up to 30 per cent greater volume (maximum 16 pallets as against 11). There are now (1981) 1,500 of these containers. A railway wagon can accommodate two 7 metre containers.

- **Tanks.** These are basically liquid-carrying cylinders with bottom and top frames for stable storage and handling. With the Conny system (used by Nynäs and Arla for petrol and milk respectively) empty tanks are returned stacked two high, leaving half the railway wagon free to house other cargo.

- **Lagab minicontainers** (base measures 2.50 × 3.40 metres) increase railway transport capacity in comparison with ISO containers since the 3.40 m width conforms to wagon width.

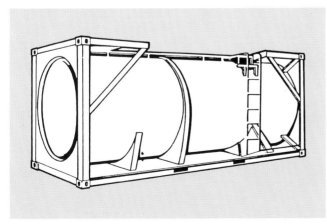

Container for liquids. The dimensions and corner castings of the end units are the same as for a standard container.

Lagab minicontainer.

Loading Lagab minicontainers from lorry to railway wagon.

Handling without pallets

Although pallets are an excellent aid to goods handling and storage they may cause drawbacks in long distance transport where high cost vehicle space has to be utilized to the maximum. Goods may be either re-loaded manually from pallet to the vehicle deck or placed on cardboard slip sheets to be pushed off the vehicle by a specially designed unit. The latter method is less common in Sweden but used frequently in the USA where lorries often cover very long distances and, consequently, the degree of vehicle space utilization is particularly important.

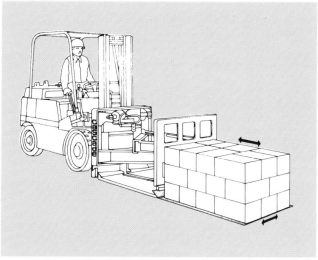

Truck with slip sheet gripper attachment.

Warehouse layouts

Storage methods - a brief summary

Goods can be stored in four ways:

- **Block stacking** on the floor or in block stacking racks.
- **In pallet racks or on shelves** where goods are placed in locations created from beams and verticals or on simple flat shelves.
- **Live stores** where goods are fed into one end of an inclined installation (equipped with roller or wheel conveyor surfaces or consisting of a metal chute) and collected at the other end.
- **Mobile, high density stores** where goods are placed on shelving, pallet racking or cantilevered racking mounted on rails. The closed up units of shelving or racking can be moved along the rails, opening up aisles where desired.

Block stacking on the floor

Block stacking means that goods are stored in stacks close to each other thus preventing access to some of the items. Normally block stacking is on the floor, but irregularly shaped or fragile units are block stacked in drive-in racking.

Block stacking on the floor provides a compact, cheap and adaptable solution requiring only a truck or overhead conveyor to handle goods. However, the strength of packing materials and the floor must meet certain requirements. The method is suitable for large volumes of stable unit loads. The range of goods stored should be small as there is no access to certain units. In practice, goods are block stacked (e.g., refrigerators) to a height of 6–7 metres.

Block stacking is a cheap and adaptable storage method for a large volume of stable unit loads representing a limited goods range.

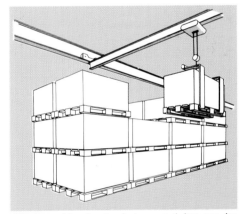

Block stacking by truck is normal but can be achieved by overhead chain conveyor.

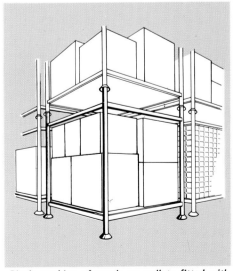

Block stacking of goods on pallets fitted with converters results in stable stacks and opportunities for picking individual items.

Block stacking in racking

Block stacking in drive-in racking is a compromise between block stacking on the floor and using conventional racking. A high degree of space utilization is achieved — valuable where rents are high, as in cold-stores — while packaging design can be kept simple. Since the unit loads are not supporting each other, access to all units is possible. On the other hand, pallet racking is fairly expensive and thus less commonly used for block stacking.

Drive-in racking for block stacking is employed for fragile or unstable goods. The truck enters the racking with its load at the correct level.

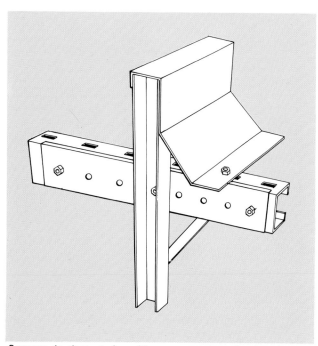

Supports in deep stack racking are sometimes fitted with pallet alignment strips.

Block stack racking equipped with inward facing supports is also used for automated high-level storage since forks find ample access beneath the pallets. In Sweden, this design is becoming popular for handling pallets on the long side in order

picking warehouses, partly because of the ease of fork access just mentioned, and partly because it provides increased picking space compared with conventional pallet racking. An interesting variant is when loads are carried on loose slip sheets resting on the supports. If loads are narrower than the opening between the supports, they can be allowed to reach heights exceeding the distance

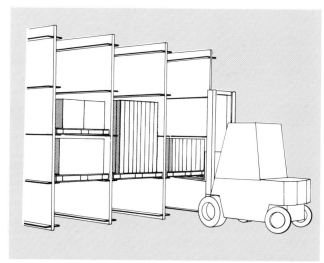

Block stack racking may also be used for loads of different heights (provided loads are narrower than the distance between supports) or to give crane forks and pickers more space.

between two levels. Thus, different load heights can be accommodated.

Pallet rack storage

Racking for storing pallet loads consists of the following components:

- Posts (verticals) in the form of rectangular hollow sections of rolled or formed sheet steel, slotted normally at 50 mm centres.
- Load supports (beams) made from channel, sigma or other sections of rolled or formed sheet metal. Beams have fasteners which engage with holes in the verticals. To prevent accidental removal, fasteners can normally be locked into position with flanges or bolts. For greater strength (mostly in high-level storage) the structure may be welded.
- End bracing, back bracing and load support bracing consist of diagonal sections bolted to posts and beams for structural rigidity. Bracings are adjustable for horizontal and vertical racking alignment.
- Sundry accessories include baseplates to distribute compressive forces on the floor, guards to prevent truck collisions from causing racking collapse, pallet stops to prevent loads sliding in too far, half-pallet inserts (dividers between horizontals), etc.

A rack end consists of a bolted assembly of two posts; two ends make a rack section. Sections are joined longitudinally to form single or double rows of racking.

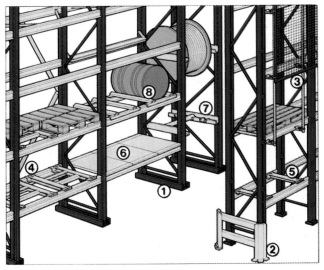

Pallet racking and accessories:
1 Baseplate for asphalt floor
2 Collision guard
3 Pallet stop and mesh panel
4 Half-pallet insert
5 Cross-member for handling of pallets on long side
6 Shelf
7 Cable drum support beam
8 Drum insert

Beams are available in standard lengths for handling pallets both on the long and short side. In both cases, there are distinctions between locations for two or three pallets (1S, 2S, 3S and 1L, 2L, respectively). The greater the width, the fewer the required number of verticals for a given number of pallets. On the other hand, more pallet locations between verticals impose greater strain on both beams and sections. The 2S solution is normally chosen as the most economical. Beams are normally designed for two 1,000 kg pallets, and a section can accommodate a load of 10 tonnes.

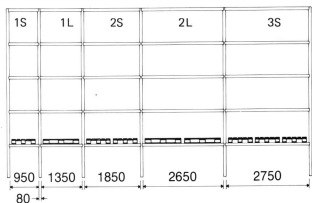

1S	1L	2S	2L	3S
950	1350	1850	2650	2750

80

Pallet racking may have different sized spaces:
1S = one pallet handled on the short side = 950 mm.
1L = one pallet handled on the long side = 1350 mm.
2S = two pallets handled on the short side = 1850 mm.
2L = two pallets handled on the long side = 2650 mm.
3S = Three pallets handled on the short side = 2750 mm.
Verticals are normally 80 × 80 mm.
Note that odd loads may be placed on the top level where there is no obstruction from posts.

Pallet racking is available in standard heights up to 7—8 metres, depending on the supplier, normally in 1.1 metre increments. Standard posts of many makes can also be built up to 10—15 metres. For high level (15—30 metre) installations, special steel sections are used, bolted or welded together.

Pallet racking is characterized by its ability to accommodate each load irrespective of the others;

this makes it suitable where storage involves a *wide range* of goods. Picking straight from stacked pallets is possible, provided there is enough space between them and the pallets above. 250—300 mm is convenient, depending on the type of goods to be picked, picking rate, picking depth (short side/long side), working height, etc. Space between pallets and verticals may be 75—100 mm.

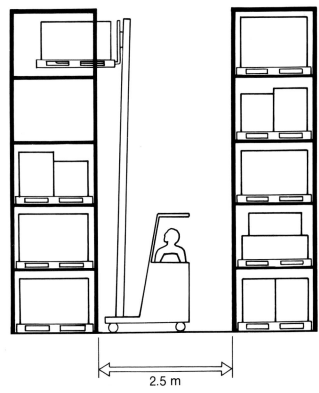

2.5 m

Aisle width requirements vary with the type of truck and pallet dimensions. Handling Euro pallets on the short side by reach truck requires an aisle 2.5 metres wide.

When planning a warehouse with pallet racking, it is essential to determine the load distribution on the floor with filled racking. Most floors can withstand 3—5 tonnes/m², equivalent to between three and five 1,000 kg pallet loads stacked vertically. Upper floors generally possess less strength, usually 0.5—1 tonne/m² for evenly distributed loads; ceiling heights limit the number of pallets. Pallet racking load capacity is indicated by permitted loads on beams and racking sections which, in accordance with the above, are 2 tonnes for a 2S beam and 10 tonnes for a section, i.e., five 1,000 kg pallet loads stacked vertically in addition to the bottom pallet resting on the floor. Strength requirements and methods for checking and testing are listed in Swedish Standard SMS 2240 (April, 1975). This also states that assembled pallet racking must not vary by more than 3 mm/m from plumb.

Racking heights are chosen on the basis of land, building and handling equipment costs. 4—6 metres is a suitable height for an ordinary unheated warehouse served by a counterbalanced truck, while a centrally heated warehouse with high capital and operating costs is designed for 6—10 metre high racking. As a guide to cost per pallet location in a medium storage installation, SEK 100 is often indicated (1980) for standard pallets in 2S compartments.

231

Shelf storage

Shelving to accommodate bins or individual goods consists of the following parts (Swedish Standard SMS 2241 design):

- Posts of formed or rolled sheet steel.
- Shelf surfaces.
- End bracing, back bracing, end panels, base panels, etc.

Units are assembled with open or closed ends which can be built up to form single or double rows. Using adjustable partitions of different height, the shelving may be divided into compartments of desired dimensions.

As a rule of thumb, shelving is used where the storage volume for an individual item does not exceed 200 litres. With larger volumes, pallet loads in pallet racking are more economical. This means it is more expensive to store goods on shelving than on pallets, despite the increased volume required for the latter.

There are two types of shelving:

- 1—3 storey manually served shelving. By using a multi-storey design, warehouse space is better utilized. Each storey may be 2.5 metres high, inclusive of space for lighting. Practical aisle width is 1000 mm.

- Mechanically served shelving. Rail-mounted or ordinary order pickers may be employed. Shelves may be up to 14 metres high, aisle widths 1000—1200 mm. This type of shelving utilizes the warehouse space efficiently, simultaneously providing increased picking capacity and better working positions in comparison with other shelving.

Shelving for storing loose or boxed items. Installations may have one or several storeys. (Mezzanines).

Rail-mounted order pickers are suitable for goods handled at low to medium rates. The picture shows a mixed solution, with high demand items on shelves served from floor level, medium demand goods handled by the order picker, and low demand goods stored in upper storeys.

Rail-mounted order pickers are particularly useful for quick, convenient retrieval from shelving.

Shelving must be able to carry the designed load with complete safety and must be provided with sufficient internal or other equivalent bracing. Normal shelves must be able to carry evenly distributed loads of 100 kg, reinforced shelves 400 kg. A single section of reinforced shelving may be loaded up to 2,400 kg.

When installing shelving, it is important to remember that the space between goods items or bins and horizontal or vertical structures should be wide enough to allow unimpeded goods retrieval. Shelves with a free internal width of, say, 950 mm are suitable for three 300 mm bins. Shelf 'headroom' is determined by the nature of the bin rims facing the picker. With a firm rim, 100 mm should be sufficient in most cases. A folding front rim requires less clearance.

Storing goods in boxes results in quick input procedures and greater possibilities for first in/ first out operation. Checking and replenishment are facilitated when bins carry item numbers. On the other hand, there are the costs of acquiring bins (SEK 10—20 each) and handling empty bins.

High density stacking

High density storage installations consist of pallet racking, stacker cranes, a transport system, control systems and the building itself. They are often automated, i.e., cranes and transport system are remotely controlled by punched card reader, keyboard or minicomputer. Installations reach heights of up to 30—35 metres; walls and roof are frequently attached directly to the pallet racking.

Insurance companies define storage installations more than 10 metres high as high-level installations (though there is a trend towards an 8 metre limit inspired by expensive warehouse fires in the 1970s). Such installations must have protection against fire in the form of sprinklers and smoke detectors. Sprinklers should be positioned at ceiling level (with at least 1 metre clearance above the top level of goods) as well as in the racking. Different classes of fire resistance (1—4) require different quantities of water, i.e, different numbers

of sprinklers. Water is supplied through vertical mains from the roof and distributed horizontally by branch pipes. High-level installations, therefore, must always have additional space for sprinkler supply piping.

High-level pallet racking can be constructed from standard verticals to heights up to 15 metres. For greater heights, special RHS (rectangular hollow section) beams are used. Strength calculations for such installations are extensive since the racking must be designed for static stresses both from stored loads and from the walls and roof of the building, including the weight of snow. To these

Pallet racking exceeding 20 metres in height is often designed to support the warehouse walls and roof. This avoids a supporting building framework and simplifies assembly. However, the racking must be designed to absorb additional stresses from snow and wind loads.

must be added dynamic stresses from stacker cranes, transferred via guide tracks at the top of the racking, and wind loads on walls.

High-level pallet racking is not always built in steel, especially in countries where steel prices are high. Instead concrete structures are created in situ or built up from prefabricated components. The latter method, in particular, provides opportunities for constructing columns with continous backs and sides, reducing the risk of any fire spreading. As a result, insurance companies view concrete high-level storage installations favourably.

High level pallet racking served by stacker crane.

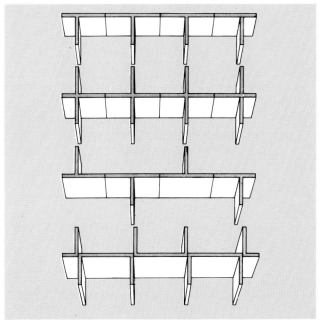

Pallet racking built from prefabricated concrete T-or U-shaped units to which shelving is fitted.

Cantilevered racking

Lengthy loads can be stored horizontally in slots, on cantilevered racking, or in vertical compartments. Cantilevered racking consists of columns with horizontal support arms. Arms are often shorter towards the top to improve access.

Goods are placed in cantilevered racking by sideloader, four-way truck or overhead stacker crane. Some steel stockholders have built automated installations with cranes controlled by punched cards and the steel housed in 6 metre cassettes.

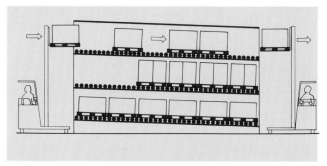

Live storage installation with roller conveyor line. Loading/unloading by truck or automated crane.

Cantilevered racking for lengthy loads, served by sideloader, four-way truck or overhead stacker crane.

Live storage

Live storage is a convenient solution for storing *large quantities* of a *few items* subject to *high turnover* rates. The method is of particular interest for goods requiring first in/first out handling, such as foodstuffs.

Live storage installations have inclined conveyor lines at several levels — wheel conveyors being suitable for boxed goods and roller conveyors for pallet loads. The rate of inclination is approximately 3 per cent, depending on goods weight and pallet condition. Installations for boxed goods can be automated, with input and output operated by pneumatic cylinder and load stops, respectively.

Live roller conveyors require some accessories such as braking rollers to restrain load speeds. This means that units which enter an empty line do not reach excessive speeds and do not cause load shifting when the pallet hits the end stop. Centrifugal type brakes are often used.

Unloading requires a separation unit which restrains upstream pallets from jamming the lowest pallet.

Loading/unloading live storage pallets is by truck or automated crane. Trucks use ordinary forks, while cranes may either use extendable forks or a powered roller table.

Live storage installations can be constructed from pallet racking components and sections of roller conveyor. Conveyor lines incorporate braking rollers, and stop and separation facilities at the output end.

For boxed goods, there are installations incorporating metal chutes fitted with remotely controlled output mechanisms. At the output end, a collecting conveyor moves discharged goods to packing or pallet loading stations. By using a pushbutton console, punched card reader or computer at the warehouse control centre to release the mechanisms, an automated order picking installation is obtained. Input is normally by hand but automated goods input is also employed.

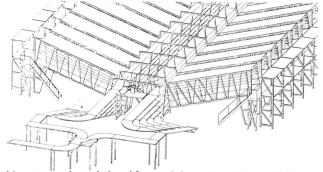

Live storage chute designed for goods in cardboard boxes. May have automated input and output.

There are other variants with fixed output arrangements, e.g., shuttle trolleys which pick up loads in their own containers from the end of the line or merely open the line to release the ordered quantity to a conveyor.

In addition to inclined chutes and conveyor lines, other live storage installations involve trolleys running on level rails. In one arrangement, trolleys — one to a line — automatically feed loads one by one along the line. In a similar arrangement, gravity-fed carriages holding pallets may have their speed augmented by an auxiliary cable drive.

Live storage is of interest where it co-exists with conventional storage methods in installations where goods have to be moved, e.g., from machining to assembly. Installations up to 10 metres high and 25 metres long have been built. The price per pallet position in a live storage installation is considerably higher than for pallet racking: SEK 700—900 depending on the size of the installation and the automation involved.

High-density storage

Low demand goods are preferably stored in high-density storage installations. These consist of mobile shelving or pallet racking in which aisles can be opened up when required.

With shelving that can be moved laterally or longitudinally, the operation is often manual since loads are not too high. Mobile pallet racking, on the other hand, runs on floor rails and is powered electrically. Mobile racking may be up to 8 metres high and 20 metres long. Air-cushion movements have also been tried.

High-density storage is suitable for archives, libraries, spare part stores, etc., where individual item retrieval rates are low, or on premises where space costs are high, such as cold-stores. A pallet position in such a high-density storage installation may cost SEK 700—900 (1980). A radio link between truck and pallet racking mechanism enables the operator to control movements by remote control to obtain aisle access.

High-density storage racking (mobile racking) requires fewer aisles and therefore increases storage capacity — as in this cold store.

Bulky goods

Bulky goods are normally stored on the floor. With odd products (such as car windscreens, tractor tyres, etc.), it is often impossible to produce sensible unit loads for mechanized handling, and simple manual methods must be accepted. Concrete conduits and telephone poles are examples of bulky goods which can be stored out-of-doors.

Goods reception and dispatch equipment

Goods reception design

Goods reception design is determined by the loading and unloading methods most suited to the goods and their subsequent handling, by structural and layout requirements and by expected goods turnover and peak utilization rates.

A useful aid to planning a warehouse is the advice and instructions contained in Swedish Standard SIS 841005, Goods Reception. Required driving space, suitable loading dock heights and headroom for different types of vehicle are among the factors covered.

The illustrations show some of the SIS recommendations for bay design and driving space.

Existing buildings may prevent the adoption of SIS recommendations. Advice and instructions about alternative measures are found in TFK Report 1980:9, 'Goods reception in a city environment'.

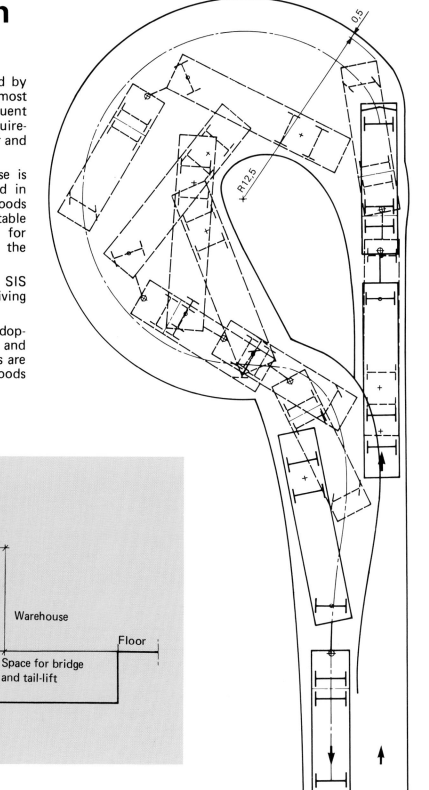

Manoeuvering space for 24 m lorry with trailer according to SIS 841005.

Bay design according to SIS 841005.

236

Loading and unloading methods

Lorries and railway wagons are loaded/unloaded in many ways (see tabular survey). It is possible to work from ground level using trucks, cranes or tail-lifts or to have a raised bay or recessed drive-way permitting hand pallet and other trucks to enter vehicles. Another solution is automated loading where the entire load is moved in a single operation onto a roller, chain or other type of conveyor.

Loading and unloading methods			
	Handling aids	**Suitable goods types**	**Supplementary equipment**
Handling from ground level	Hand pallet truck	Pallet loads	Tail-lift
	Sack truck	Loose general cargo	Tail-lift + manual handling
	Overhead rail conveyor or gantry crane	Lengthy or heavy loads	
	Vehicle mounted crane	Pallet loads, lengthy loads	
	Counterbalanced or reach truck	Pallet loads	Portable loading ramp or access through side
	Belt conveyor	Cardboard boxes, sacks	
Bay handling	Sack truck	General cargo, cardboard and other boxes	Bay, lifting platform, dock seal
	Hand pallet truck	Pallet loads	
	Counterbalanced truck Powered stacker Reach truck Roller conveyor Chain conveyor Slat conveyor	Pallet loads	

Unloading from ground level is facilitated if goods are carried on pallets and accessible through side openings.

Unloading by forklift from ground level (counterbalanced or reach truck) can be a very efficient method. Its attractions are enhanced if all goods are carried on pallets or fitted with spacers and if lorries/rail wagons have fully opening sides providing access to cargoes.

If the lorry or rail wagon has a limited opening facility (at the rear, or via a small side door) handling aids will be needed. The same applies if goods are not carried on pallets or are of a type unsuited to pallet handling. A bay is then the natural solution. There are several types: end unloading, side unloading and staggered. A staggered bay may be an attractive compromise since goods can be handled from the rear and the side manoeuvering space requirement is kept to a minimum.

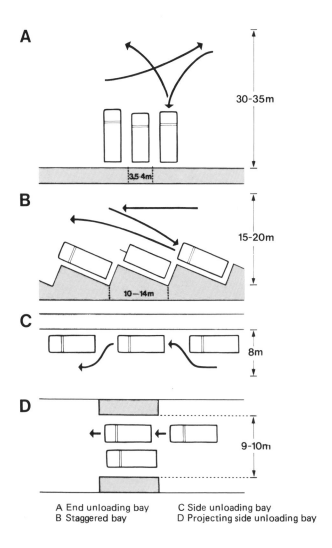

A End unloading bay   C Side unloading bay
B Staggered bay   D Projecting side unloading bay

Portable loading ramp.

In practice, various simple measures can compensate for variations in height. A variety of bay heights is one answer; with very low vehicles detachable ramps or blocks may be used.

Small portable ramps may solve problems caused by varying lorry deck heights.

Tail-lifts also compensate for different bay heights and are a standard solution to goods reception problems in the grocery/household goods trade.

Varying lorry deck heights create problems
Most goods reception areas must take account of the fact that they will be used by many different types of vehicle, with deck heights which may vary between 0.7 and 1.5 metres. The key to bay handling, therefore, is to achieve appropriate level adjustment without employing troublesome gradients and sills.

The most common solution is to equip the bay with manually or hydraulically operated dock bridges.

A less common alternative is to supplement the bay (or perhaps replace it entirely) with a lifting platform (see section 'Elevators'). Portable loading ramps may also be used.

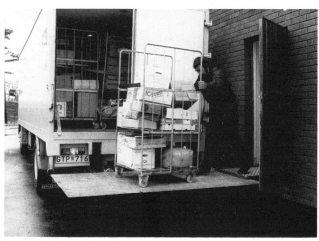

Distribution lorry with tail-lift.

Working environment and safety

To a great extent goods are still received in outdoor installations exposed to draughts, low temperatures, rain and other vagaries of climate. Because of increased demands for acceptable working conditions and safety in goods reception — as well as everywhere else — these aspects are now considered during the planning of an installation.

Indoor loading/unloading is the best solution and some new distribution centres, where handling rates are high, do incorporate this.

A cheaper and more usual alternative is to load/unload vehicles from the rear through specially designed doors. To create the best possible environment and to keep heat losses to a minimum doors may be equipped with flexible dock seals against which vehicles reverse.

However, this is not entirely satisfactory compared with indoor handling as seals may not mate with every type of vehicle and could interfere with handling operations.

Handling efficiency and thermal economy are affected by the nature of the doors used in goods receptions. Guidance for choosing industrial doors is given in TFK Publication No. 80, 'Door handbook'.

Handling goods on open bays which lack separate loading doors — still the most usual method — puts operators at risk from falling unless special measures are taken. Advice and instructions on suitable bay safety arrangements are given in TFK Report 1977:6.

Bay arrangements at a wholesaler's with seals and bridges.

Bridges

There is a vast range of dock bridges, from simple portable constructions in formed sheet metal to large pneumatically or hydraulically operated permanent structures.

Mobile dock bridges.

Mobile dock bridges (see picture) which can be moved manually are suitable where goods reception involves only a few vehicles and where a more expensive installation would be uneconomical. The simplest versions are carried by hand to the loading point; this limits both size and capacity. On the basis of a maximum acceptable gradient of 10°, such small bridges will accommodate approximately 100 mm level differences, but this may often prove insufficient as lorry deck heights can vary by between 800 and 1,500 mm, i.e., by up to 700 mm.

239

Slightly higher capacities are achieved with semi-permanent dock bridges which are mounted at the edge of the bay. They, too, are manually positioned which limits their levelling abilities to 200—300 mm.

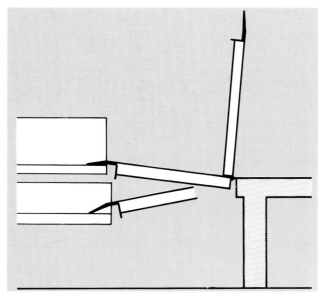

Semi-permanent dock bridge.

Much higher capacity is achieved with heavier detachable bridges positioned by hand pallet truck, powered truck or special trolley. Length is up to 3 metres and level differences up to 500 mm can be accommodated.

The natural solution for terminals and goods receptions where there is heavy traffic is the permanent, mechanized bridge. This is often hydraulically operated but may be spring-loaded or counterbalanced.

Spring-loaded bridges are normally positioned somewhat faster than hydraulic models although the latter provide opportunities for a higher degree of automation. The most sophisticated

models automatically take up the correct loading position after a vehicle has reversed into place, returning to the parked position when the vehicle leaves the bay.

The bottom left figure shows a hydraulically operated dock bridge. When the bridge is not in use its far end can be folded away with a handle to give some protection against falls. It is also fitted with articulated flaps which provide some angular adjustment between bay and lorries. Length is up to 4 metres. Height differences of 500—600 mm and loads up to 7 tonnes can be accommodated. This type of bridge is often recessed in the bay itself — a natural solution if a bridge was planned from the start.

With existing bays a non-recessed bridge may prove simpler and cheaper.

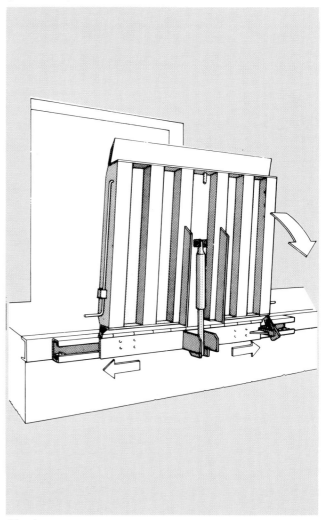

Simple spring-loaded dock bridge mounted directly on an existing bay.

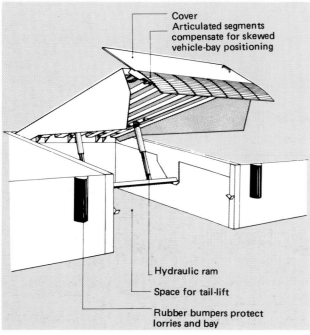

Cover
Articulated segments compensate for skewed vehicle-bay positioning

Hydraulic ram

Space for tail-lift

Rubber bumpers protect lorries and bay

Hydraulically operated dock bridge.

These, too, are available in hydraulically operated versions. They are primarily used for lighter loads (under 3 tonnes) and offer moderate levelling facilities (less than 400 mm). Handling on them is less safe than with recessed bridges since lorries cannot reverse right up to the bay.

When installing a dock bridge the need for sufficient tail-lift space should also be taken into account.

EXAMPLES OF LOADING BRIDGE SPECIFICATIONS

Bridge type	Max. load kg	Length mm	Levelling mm	Dead-weight kg	Handling method	Material	Approx. price SEK (1980)
Mobile manual bridges	500–1,000 2,500–4,000 2,000–4,000	600–1,200 400–800 1,200–2,100	80–200 25–150 140–320	15–30 24–40 45–100	Lifted by hand Lifted by hand Moved by hand	Aluminium Aluminium Aluminium	510–1,300 900–3,000 2,000–6,000
Mobile non-manual bridges	1,000–4,000	1,900–3,000	210–625		Positioned by special lifters or forklift trucks	Steel	3,000–8,000
Semi-permanent bridges	2,000–4,000 3,000–6,000	500–1,100 1,500	80–160 220	25–60	Raised by hand Manually operated (spring-loaded)	Aluminium Steel	1,000–2,700 4,500–7,500
Permanent spring-loaded bridges	3,000–9,000	750–1,500	125–250		Manually operated	Steel	3,200–4,500
Hydraulically operated bridges	3,000–9,000	1,300–5,000	300–650			Steel	13,000–25,000

Loading ramps

Loading ramps enable trucks and other handling aids to enter lorries, rail wagons and containers without using a dock bridge. They may be suitable for goods receptions where unloading is mainly from ground level and there is only an occasional requirement to enter a vehicle with handling equipment. Loading ramps may be permanent or mobile.

Loading ramp used to load and unload a container parked on the ground.

Dock seals

A dock seal is generally made from some flexible material such as reinforced plastic or rubber mounted on a steel frame. It is fitted round a loading door, and lorries seal the opening by reversing up against it.

Dock seals may be used when loading/unloading refrigerated and chilled products, electronics and other products vulnerable to temperature changes or to the weather.

Goods may not require any such protection, but the dock seal may still be justified in that it protects the workforce.

A dock seal may consist of soft cushions, e.g., filled with aerated plastic, which are deformed by the reversing lorry. Provided the lorry dimensions are fairly uniform, this type of seal is effective and is used predominantly in connection with handling refrigerated or chilled products.

Segmental type dock seal.

The second type usually consists of plastic fabric or rubber segments at the sides of the door and above it. They are bent inwards by the vehicle and, to improve efficiency, are often spring-loaded. The fabric/segments are mounted on a steel or wooden frame which is either immovable or flexible. In the latter case, it yields to pressure from reversing vehicles. Such dock seals are used in places such as general cargo terminals.

Cushion type dock seal. The cushions are supplemented by overlapping segments to reduce wear caused by vehicles.

Equipment for assembling and securing loads

Producers, wholesalers and operators of distribution terminals are becoming more aware of the importance of assembling goods in rational units for dispatch (see also section 'Load carriers'). Units should be adapted to the size of the order and to the handling aids available. Depending on the nature of the goods and on the pattern of orders units may take many different forms.

For small goods of various shapes and sizes, a closed container may be the best solution, i.e., a plastic bin, a pallet with collar(s) or a box pallet, depending on goods quantity and consignment size.

More uniform and regularly shaped goods (such as standard size cardboard boxes and sacks) may be loaded on pallets without collars or into box pallets. For more secure handling, goods may be held together by strapping, plastic film, etc.

Some goods can be built up into unit loads — suitable for handling by forklift — without the need for pallets.

Goods may be assembled into easily transported units by hand, by various degrees of mechanization or fully automatically. This section examines the various equipment used.

The table surveys available equipment in relation to unit loads and goods types.

Unit loads that do not depend on pallets, yet may be handled by forklift (e.g., sacks and cardboard boxes) can be built up with shrink film. However, they cannot be stacked in conventional racking.

Aids to assembling and securing loads		
Unit loads	Equipment	Suitable goods
Pallet (without sides)	Palletiser (semi- or fully automatic	Cardboard boxes, sacks, bins, slabs
	Pick-and-place device	Cardboard boxes, plastic bins
	Bin stacker	Plastic or wooden bins
	Glueing equipment	Cardboard boxes, sacks
	Strapping equipment Shrink film equipment Stretch film equipment Netting	Cardboard boxes, bins, wooden boxes, etc.
Roll pallet (pallet on wheels, with two sides)	Pick-and-place device	Bins, cardboard boxes
Box pallet	Automatic box pallet filler	Cardboard boxes
Container	Automatic box pallet filler	Cardboard boxes
Non-pallet unit	Pattern assembler Shrink film equipment Stretch film equipment Strapping equipment	Sacks, cardboard boxes

Palletisers

Palletisers are fully or semi-automatic machines for placing goods on pallets. The goods may be cardboard boxes, plastic or wooden bins, sacks, slabs, etc.

Palletiser for crates.

Machine for stacking wooden bins on standard pallets.

In a **semi-automatic palletiser** individual units are assembled manually in a given pattern, and this layer is then transferred mechanically to a pallet.

In a **fully automatic palletiser** pattern assembly is mechanical. The machine may be designed as shown below.

Pattern assembler (1) places goods items in the desired pattern.

Layer assembler (2) collects a layer of goods items and places it on a loading board or on forks.

Loader (4) pushes the goods layer into place above a pallet which has been fed to the loading position from a **pallet magazine** (3).

Loading board (or forks) is retracted, so that the layer rests on the pallet.

Discharger (5) removes the completed pallet load.

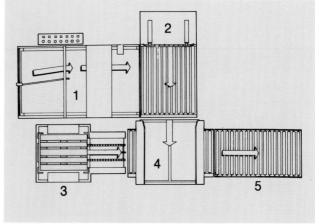

Principle of a fully automatic palletiser.

Charging is based on either of two principles. Usually the pallet is positioned in a lift which is lowered one step for each layer loaded.

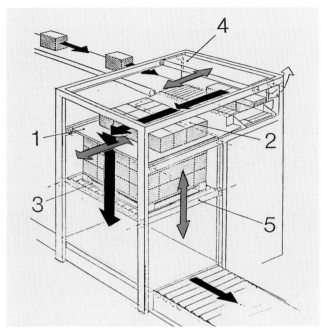

Palletiser lift lowers the pallet as each layer is added.

In other palletisers, pallets remain in the same position. Goods are loaded on to the pallet, layer by layer, by a combined clamping and lifting unit. The advantage is that the lifting machinery can be designed to lift only one layer (e.g., cardboard boxes) at a time. The principle can be used for both fully and semi-automatic machines.

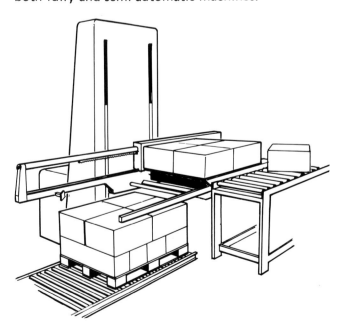

Palletiser without lift. The machine lifts, transfers and lowers one layer of goods at a time, depositing it on the pallet. The lifting mechanism is designed to handle only one layer. Pattern assembly may be manual (as in the picture) or mechanical.

For sacks, there is a special type of palletiser which operates without lift. Sacks are carried by forks which move in vertical tracks and are deposited layer by layer on the stationary pallets as the forks are retracted.

Sack palletiser. Sack layers are transferred by forks which move in vertical tracks.

A sack compressor is often used in conjunction with a palletiser to achieve uniform and even patterns and layers on pallets.

Bin stackers

Goods picked from a wholesaler's or producer's warehouse are often dispatched in plastic or wooden bins. Logically, these can be combined into larger units, e.g., single stacks for handling manually or by special trucks (see section 'Carts and trolleys'). Alternatively, a larger number of bins may be placed on a standard pallet or roll pallet. Special bin stackers have been developed for the purpose.

For stacking bottle crates, etc., machines are available which transfer a complete layer (6 or 9 crates) at a time to the pallet.

Pick-and-place equipment

So-called pick-and-place units are comparatively new devices which pick and assemble goods in one cycle.

Machine for stacking 40 × 60 cm bins. When a stack reaches the desired height it is automatically placed on a standard pallet or roll pallet.

The pick-and-place unit runs on rails between two rows of pallet loads of goods.

The photograph shows such a device handling cardboard boxes. The machine moves on rails between stacks of pallets. The operator drives to the pallet location, lifts boxes off the pallet by suction gripper and places them on an empty roll pallet in front of him. Note that a pick-and-place unit is not an alternative to a palletiser. The latter automatically or semi-automatically transfers uniform goods items, picked earlier, to pallets.

Pick-and-place units are suitable for picking goods from a limited range and depositing them on a pallet. The range of goods may be extended if the machine's movements are extended longitudinally and/or if the machine is designed to accommodate several transverse pallet positions. In practice, however, it is likely to be limited to a few dozen products. Equipping the lifting device with different types of grippers permits the handling of different types of goods.

Pick-and-place unit with suction gripper may be used for cardboard boxes and other items which have fairly even top surfaces that are air-tight, e.g., shrink film covered containers.

For beer crate handling a pick-and-place unit uses mechanical grippers capable of holding five-crate stacks.

Filling box pallets and containers

Box pallets, with or without wheels, are used more and more to transport general cargo (parcels, packages). They are normally filled by hand. A sorting installation may supply goods to an accumulating roller conveyor from which items are lifted manually and placed in roll pallets. Where parcels are robust, they can obviously be tipped straight into sacks or other containers. Where there is a risk of damage or of unsatisfactory filling, a special filling machine may be used.

The photograph shows a machine for filling roll pallets. It should be used in combination with automatic sorting methods. Items slide into the container in which a platform descends successively as the container fills up. This reduces the drop distance to a minimum. When the container is nearly full the platform re-emerges at one end of the machine.

Pallet inverters

In goods assembly and dispatch, pallets sometimes have to be swapped, perhaps because of the need to replace a high quality in-house pallet with a simpler and cheaper pallet. Handling food is a typical application where special hygienic pallets are used internally.

A pallet may also be changed because it is defective or because it needs to be swapped for two half-pallets. Special pallet inverters (or pallet changers) perform this job.

Rotary drum pallet inverter.

Pallet inverters may be of a rotary drum design like the one pictured above. To change a pallet the unit load is placed on the pallet support surface. The exchange pallet is positioned on top of the load, and a pressure plate is lowered to hold the load securely while the drum is rotated through 180°.

Pallet inverter which permits handling by hand pallet truck.

The figure shows a different design in which the pallet inverter is a clamping unit. It can be used for 180° or 90° rotation followed by return to the initial position. With rotation through 180°, loads cannot be positioned and retrieved from the same side (as opposed to loads in a drum inverter). On the other hand, pallet loads can be handled by hand pallet truck at floor level.

Load security

A unit load built up on a pallet with collars or in a box pallet is generally sufficiently stable to withstand transport without need for additional support. But where goods are assembled on a pallet with no supports it may be necessary to secure the load to ensure it remains unitized throughout loading/ unloading, transport and any subsequent terminal handling.

Pallet loads may be secured in many different ways.

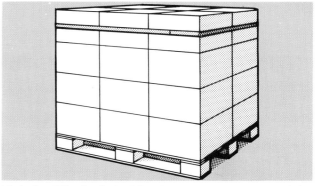

Limited stability is obtained by simply placing a strap round the top goods layer. The strap may be a rubber band, adhesive tape, string or plastic strapping tensioned by a special tool (see section 'Strapping').

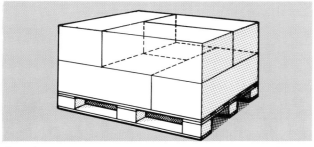

Sometimes, acceptable stability is obtained simply by positioning goods in a bond pattern.

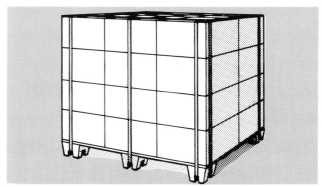

By strapping spacers under the goods, possibly with a thin intermediate board, a 'pallet load' is produced without a proper pallet. This offers cost advantages but requires relatively stable load components.

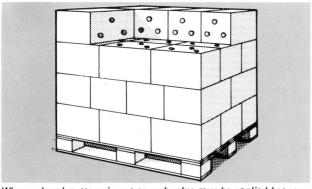

When a bond pattern is not enough, glue may be applied between items (see figure) or friction inserts added between layers.

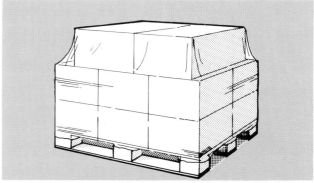

Shrink or stretch film. Wrapping entire pallet loads with thin, prestressed plastic film is now very common. Provided the goods are not too heavy, the technique is satisfactory and it can also provide good weather protection. Equipment for applying the two types of film — shrink and stretch — is described in a special section.

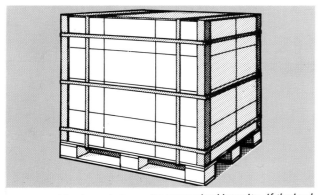

Vertical and horizontal straps give greater load integrity. If the load is also fitted with edge protectors and, possibly, a firm cover, strap tension may be increased without straps cutting into the goods. Straps may be plastic or steel (see next page).

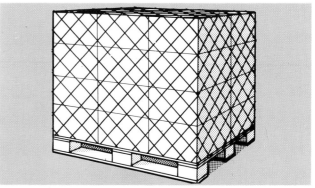

Pallet loads may also be secured with netting made from conventional non-stretch fibre (usual for aircraft pallets) or elastic pretensioned netting which is wound round the pallet.

Strapping

Strapping, a well-proven method of holding goods and unit loads together, is simple and comparatively cheap and can be applied with either hand tools or more complex automatic machines.

However, strapping only suits goods which have a fairly regular shape and which are not too soft. Strapping tools often need a fairly hard surface to be of use. Because straps can damage goods at the edges, edge protectors may be needed.

Examples of some typical applications:
- Wood-based unit loads such as paper, pulp and sawn timber.
- Unit loads of metal products such as piping, beams, sections and sheet.
- Heavy cardboard cartons on pallets.
- Unit loads on pallets protected by frames and covers.

Shrink or stretch film wrapping may be better for vulnerable and lightweight goods, especially if they must also be protected from moisture and dust.

Strapping materials

Originally straps were made of steel. Lately, synthetic materials have become more common — at least for lightweight goods. A wide range of plastics is available, offering different grades of elasticity and strength and different joining methods.

Choice of strapping depends on the nature of the goods. Steel strapping is not really suitable for loads which bed down, often as a result of vibration during transport. Nylon with its pronounced elasticity is more suitable.

For steel products and goods with sharp edges, steel strapping may be the best solution since plastic strapping could be damaged or even fail through abrasion.

For a fairly stable load which is unlikely to change shape, polyester strapping may be preferable. It retains pre-stressing better than other plastics.

The table shows various strapping materials and their most important properties.

Properties of strapping materials				
Material	Tensile strength N/mm^2	Elongation %	Typical dimensions mm	Special properties
Cold rolled low carbon steel	700	2 approx.	16 × 0.5 19 × 0.6	Low elasticity. Suitable for light applications
Cold rolled medium carbon steel	800	5–8	16 × 0.4 19 × 0.5	Elongation starts long before fracture.
Hot rolled high-tensile steel	1,000	10–12	19 × 0.8 32 × 1.27	For heavy loads
Polyester	600	6–8	13 × 0.5 16 × 0.5	Retains prestressing better than other plastics.
Nylon	450	18	13 × 0.4	Highest recovery rate after elongation. Superior resistance to sunlight
Polypropylene	350	14	13 × 0.4	Lowest weight and cost compared with other materials.

Strapping equipment

Equipment for applying both steel and plastic strapping ranges from simple hand tools to large automatic machines for multi-directional strapping operations.

The table shows available equipment in relation to strapping methods.

Strapping method	Equipment	Sealing method	Strapping material
Manual without tools	Self-locking buckle		Plastic, rayon
Manual with simple tools	Tensioner, manual or pneumatic	 Locking sleeve, punched Locking sleeve, crimped Punching directly in strapping	Steel or plastic
	Crimping tongs, manual or pneumatic		
Manual with combination tool	Combination tool for tensioning, punching and cutting, manual or pneumatic	See above	Steel or plastic
	Combination tool for tensioning, welding and cutting	 Friction welding	Plastic only (provides superior strength)
		Thermal welding	Plastic only
Semi-automatic strapping	Single or multi-strand strap feeder plus pneumatic combination tool	See above	Steel or plastic
Fully automatic strapping	Automatic machines for vertical strapping of rectangular items Automatic machines for vertical strapping of circular items Automatic machines for horizontal strapping Automatic machines for combined horizontal and vertical strapping	Same as above plus twisting (thin steel straps or wire only)	

Manually applied plastic straps employ a simple self-locking buckle (see figure in table). An advantage in comparison with other strapping methods is the possibility of post-tensioning. However, the method is suitable only where slight tensioning is required.

Greater tension demands separate tools for tensioning, joining and cutting straps or combined tools for two or three functions.

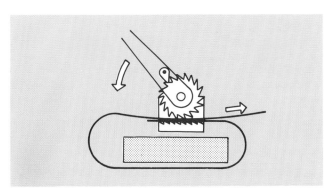

Tensioner with support plate and friction wheel. Tensions unlimited strap lengths but requires loads with fairly flat surfaces and structural resistance.

Tensioners are available in two principal designs.

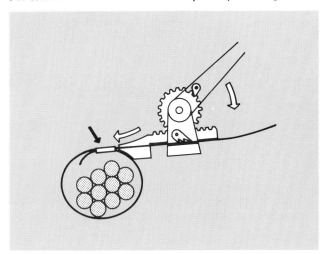

Tensioner without support plate (push type). Is used for strapping irregularly shaped items. The locking sleeve acts as a counter support.

Hand tools for strapping

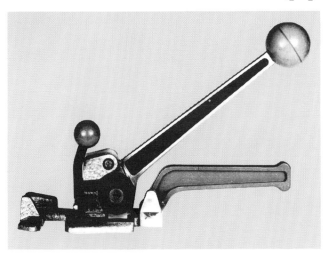

Manual push type tensioner.

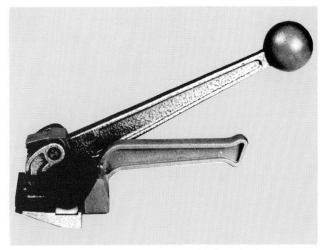

Manual tensioner with support plate and friction wheel.

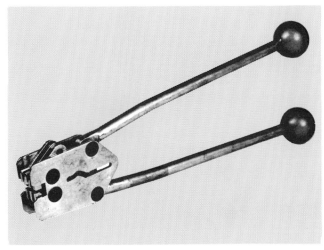

Sealing tongs for locking sleeves.

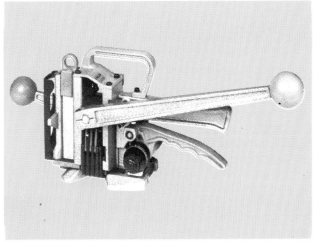

Manual combination tool for tensioning, sealing and cutting, with magazine for locking sleeves.

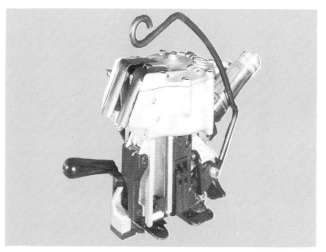

Pneumatic combination tool for tensioning, sealing and cutting.

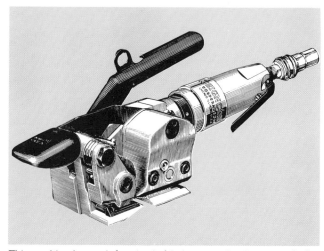

This combination tool, for plastic friction strapping, is pneumatically operated but there are also electrically operated versions for light duty strapping.

For strapping large volumes of goods, automatic or semi-automatic strapping machines may be preferable.

A semi-automatic strapper has one or several arms feeding one or several lengths of strapping round the goods. Tensioning and sealing is then by pneumatic or electric hand tool.

The best solution for large volumes of uniform goods may be fully automatic strapping.

An automatic strapper handles feeding, tensioning, sealing and cutting without manual intervention. The machine may be pushbutton controlled or may form part of an automated packaging line.

The pictures below show a few examples:

Semi-automatic strapping. Strap feeder for applying plastic strapping to pallet loads.

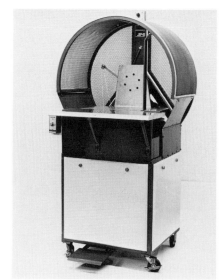

This machine wraps twine once or twice round the item, using a rotating arm; the twine is then tied automatically. An attachment allows cross-wrapping.

The photograph shows a strapping station with a strap feeder for 13 × 0.5 mm plastic. Sealing is by manual combination tool. Note that the load has been fitted with edge protectors before strapping. Semi-automatic strapping stations are available in a great many different designs for vertical or horizontal strapping or both. In horizontal strapping, some strap feeders encircle the load. Special designs can handle sheet steel coils and paper reels.

Automatic application of plastic strapping to wooden bins and cardboard boxes. Sealing is by friction or thermal welding. Capacity is up to 25 complete cycles per minute.

Automatic (pushbutton controlled) steel strapping of pallet loads. The loads, consisting of paper, are protected by wooden board covers.

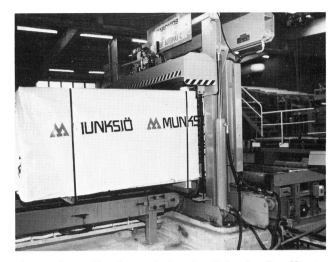

Automatic machine for steel strapping timber bundles. May be equipped to position corner protectors and spacers (for forklift handling) automatically. Such machines usually provide a unit load which can be lifted by its strapping harness.

EXAMPLES OF STRAPPING EQUIPMENT SPECIFICATIONS

Equipment type	Strapping material	Sealing method	Tensioning force N	Strapping cycles/ min	Approx. price (1980) SEK
Manual tensioner	Plastic		100–5,000	1–5	700
Manual tensioner with feed wheel	Steel		2,000–7,000	1–5	1,500
Manual tensioner with friction wheel	Steel		2,000–10,000	1–5	700
Pneumatic tensioner	Steel		2,000–40,000	2–6	6,000–8,000
Manual combination tool	Steel	Punching Locking sleeve	2,000–10,000	2–6	2,500
Pneumatic combination tool	Plastic	Locking sleeve	100–5,000	2–6	6,000–8,000
Pneumatic combination tool	Steel	Punching Locking sleeve	2,000–15,000	2–6	6,000–9,000
Electric combination tool	Plastic	Thermal welding	100–5,000	2–6	3,000–5,000
Strap feeder	Steel Plastic			1–3	15,000–30,000
Automatic strapper for light loads	Plastic	Locking sleeve Friction welding Thermal welding	10–1,000	20–40	30,000–50,000
Medium fast automatic strapper for light loads	Plastic	Locking sleeve Thermal welding	10–1,000	10–20	30,000–50,000
Automatic strapper for newspaper bundles	Plastic	Thermal welding	10–1,000	25–35	60,000
Heavy duty automatic steel strapping machine	Steel	Locking sleeve	10,000–40,000	1–5	150,000–500,000

Shrink and stretch film wrapping

Securing loads with shrink and stretch film wrapping has lately become more and more popular. For example, it may be employed with vulnerable items which cannot be strapped because they cannot tolerate the strain imposed, or with items which are too irregularly shaped to be strapped. Shrink or stretch film wrapping may be superior for large pallet loads consisting of a great number of small items since film alone replaces many straps plus edge protectors and a top cover.

However, film is no substitute where heavier goods demand greater holding power.

Shrink film wrapping employs polythene film shrink stressed by a special stressing and cooling process. The film is normally wrapped round the load and welded. Alternatively, ready-made plastic hoods may be slipped over the load. When heat is applied (as hot air or by infra-red radiation) the built-in stresses are released, shrinking the film so that it stabilizes the load.

The method is less suitable if goods are heat-sensitive. Also, it is comparatively wasteful of energy.

Stretch film wrapping, therefore, has been developed as an alternative. The film (polythene or PVC) is tensioned and extended mechanically as it is wrapped round the load. The advantages and disadvantages of both methods are listed in the table below.

Comparison of characteristics	
Shrink film	**Stretch film**
Shapes well	Provides greater holding power
Simple to seal completely	Laid mainly in one direction
Holds goods together in all directions	Offers greater capacity
'Curl over' effect covers ends	Requires no heating
Creates more even surfaces	Often requires little space
Rapid cycle	Effective with several layers of thin film
Film manufacture usually simple	More suitable for covered transport
More gentle to fragile goods	No lamination problems
	Less energy consumption

Film materials

The most common film material is polythene (PE), available in many grades and usually rated on the basis of density (weight per unit volume).

PE grade	Density kg/m^3
LDPE (low density)	910–925
MDPE (medium density)	926–940
HDPE (high density)	941–965

LDPE is the common grade for shrink and stretch film, sometimes combined with a co-polymer such as ethylene vinylacetate (EVA) providing rubbery flexibility and a low softening temperature. A more recent constituent for stretch film wrapping is PVC containing a plasticizing ingredient which makes the film attract static electricity so that it sticks to itself, eliminating the need for welding. It is tougher than polythene and can be stretched considerably more. Also, it recovers its initial shape better from stretching and is not as prone to damage. Perforations are not enlarged to the same extent as with PE plastics. It costs about twice as much, but its greater strength and, consequently, its reduced rate of consumption (1980) may compensate for the higher price.

Shrink film wrapping equipment

In the simplest application, a ready-made film hood is slipped over the load and the film is shrunk by a hand-held **shrink gun**. Guns usually operate on LP gas but there are electric versions. The time taken to shrink a wrapped pallet load is about

Shrink film wrapping by pallet hood and shrink gun.

two minutes.

Gun heating may carry some risk to operators although guns normally have a 'dead man's grip' which turns off the heat when released. This method consumes more film than shrink ovens and tunnels.

It is possible to mount one or several heat sources vertically on stands and let goods rotate in front of them on a turntable. Alternatively pallet loads may pass between two stands fitted with heaters. This produces shrinking on only two of the four sides.

Yet another alternative is a mobile shrink assembly consisting of a vertical heating element mounted on wheels. It may be manually propelled round the load, it may be self-propelled or it may possibly be automated.

Capacity is considerably greater than that of a shrink gun and pallet loads up to 1750 mm high can be accommodated.

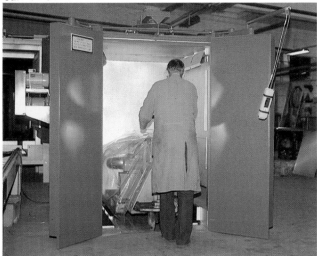

Shrink oven with automatically operated doors.

With large throughputs, the most commonly used method is some kind of **shrink oven**. The simplest version has a door on one side through which loads are carried by forklift or hand pallet truck. Such ovens are usually electrically heated. Film, 0.05 mm thick, requires heat shrinking for about a minute.

Two different design principles provide a higher degree of automation.

One, less used in Sweden, is an oven which is lowered over the load. It can be combined with a continuous pallet conveyor to form a complete packaging line. Capacity is normally 30–60 pallets per hour.

The other, more common, design is a shrink tunnel consisting of an oven with doors at each end, fed by a chain or slat conveyor.

Shrink tunnel with chain conveyor.

The load is automatically positioned correctly in the tunnel and the doors are closed. When a pre-set time has elapsed the discharge doors are opened, and the processed load is moved out for cooling. As one load is discharged, the next pallet load is conveyed into the tunnel and the cycle is repeated. Capacity may be up to 120 pallets an hour. Heat output varies between 80 and 120 kW; in other words, the method consumes considerable amounts of energy. However, tunnels may be designed to switch down and reduce the heat output by half as long as the pre-set temperature is maintained.

To obtain a fully automatic packaging line, a wrapping unit may precede the tunnel. The wrapper may have a device which shapes the plastic over the load and a welding station for sealing. Welding is at top of the load or on the side. Side welding offers more effective holding round the load.

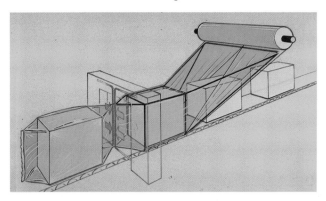

Pallet load wrapper with vertical welding at leading and trailing ends.

Complete shrink film wrapping line consisting of an automatic wrapper and a shrink tunnel.

Stretch film wrapping equipment

Stretch film wrapping, introduced later than shrink film wrapping, seems to be gaining in popularity. This is due, among other things, to lower capital costs, lower energy consumption and lower space requirements (see also 'Comparison of characteristics').

Stretch film wrapping is based on one of two techniques. The **push-through technique** by which goods are conveyed against a mechanically held curtain of film which then stretches. The film is weld sealed while tension is maintained round the load. Suitable for automatic machines with capacities of 20—100 pallets an hour.

sufficient the film is united by welding or self-adhesion. Film can be made self-adhesive by additives such as polyvinyl acetate or by surface treatment. Full width winding is differentiated from spiral winding. Winding is employed mainly in semi-automatic machines with capacities up to 25 pallets an hour.

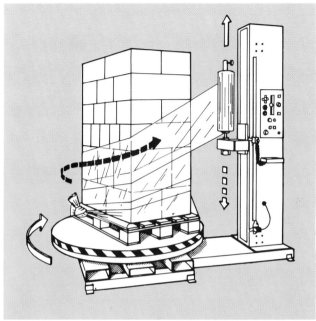

Spirally wound stretch film system. The machine is particularly suitable for varying heights of pallet load and may automatically adjust to the correct wrap height. Some machines can manage both spiral and full width winding.

Stretch film wrapping is also possible by the simpler method of moving some kind of film reel under tension on a holder round a stationary load.

The simplest aid, used for entirely manual winding, is a hand-held winder with PVC stretch film.

Fully automatic stretch film wrapper using the push through technique.

With **winding technique,** goods are positioned on a rotary table and wound with film restrained at the supply end. When the wrapping thickness is

A reel of film held on a mobile stand facilitates entirely manual, full width winding at a very small capital cost.

Powered mobile stretch film wrapping machines are also available.

Stretch film wrapping robot. When the shrink film has been attached to the pallet, the robot moves itself round the stationary load, using a wheel sensor. Three or four spirally wound turns are applied in about 2 minutes.

EXAMPLES OF STRETCH AND SHRINK FILM WRAPPING EQUIPMENT SPECIFICATIONS

Equipment type	Power input kW	Units handled per hour	Approx. cost (1980) SEK
Shrink gun	6	7—10	4,000—5,000
Shrink oven	40	25	45,000—55,000
Shrink tunnel	80	60	75,000
Spiral winding stretch film wrapper	2	20—30	30,000—75,000
Stretch film wrapping robot	0.3	20—30	

Case studies

PART 3

This section describes the way different companies have solved their materials handling problems.

An attempt will be made to show how and why the equipment was chosen, starting from the special circumstances pertaining to each company. The examples should not, however, be regarded as recommendations: they simply indicate how practical choices were made. It is left to readers to judge whether or not the choices were the right ones.

Assembling diesel engines

Saab Scania, Södertälje

Goods reception in the engineering industry

Volvo-BM, Eskilstuna

Picking and sorting merchandise

NK-Ahléns, Jordbro

Moving pallet loads in a furniture warehouse

IKEA, Älmbult

Order picking using trucks

ASSA-Stenman, Eskilstuna
ICA, Västerås

SKF's Gothenburg warehouse for the Scandinavian region

Assembling diesel engines

Diesel engine being assembled.

The situation

At Saab-Scania in Södertälje, diesel engine assembly (involving three main types of engine) was reorganized. One restricted and one unrestricted assembly line were replaced by a new system with three unrestricted lines, one for each engine type.

A mechanized warehouse for approximately 700 diesel engines, integrated with assembly and testing, was included as well as a mechanized system for handling engines between assembly, storage and testing.

A survey of the materials supply system and the engine load carriers/assembly fixtures was carried out.

Available floor area (150 × 40 m = 6,000 m²) in an existing building was set aside for engine assembly. Roof height is 6.6 m, columns are placed at 12 × 24 m centres and the floor covering is 5 cm thick asphalt tiles. The premises are intersected by three 2.5 m wide ventilation ducts, each having a 2.5 × 3 m steel access door for service and maintenance of fans.

The production system was to be reorganized during normal working and the time schedule allowed six months for completion.

261

The table below shows the subfunctions covered by the materials supply system and the types of equipment chosen for them.

	Subfunction	Equipment chosen
1	Transport between assembly stations.	Unpowered roller way to support load carrier/ assembly fixture.
2	Accumulation before and during assembly	Powered accumulating roller conveyor.
3	Re-routing from assembly line	Automatic lifting platform with powered roller conveyor.
4	Transport to warehouse	Automatically guided vehicle.
5	Stores input	Minicomputer controlled crane.
6	Storage	Live store.
7	Retrieval from store	Minicomputer controlled crane.
8	Transport to testing, adjustment, dispatch	Roller conveyor, forklift truck.

Diagram of essential subfunctions.

Load carriers /assembly fixtures

Changing over to the new assembly and handling system made it possible to rationalize transport and assembly fixtures. Instead of, as before, using wheeled vehicles and repositioning items on various fixtures, a uniform fixture was introduced to fit all engine types and to accompany each engine through the entire production cycle from assembly to installation in the chassis. (Engines remain on their fixtures also during testing.)

Material supply

The materials supply system was changed at the same time as the production system. Previous assembly on restricted lines involved supplying components in batches. Group assembly requires a different supply system and the method now used is termed 'uniform supply'. Pallet loads unique to one item are placed in racking along both sides of the assembly lines. Buffer stocks and variants to the main components are stored in the top three storeys of racking. Behind the racking are aisles for replenishing stocks.

Layout

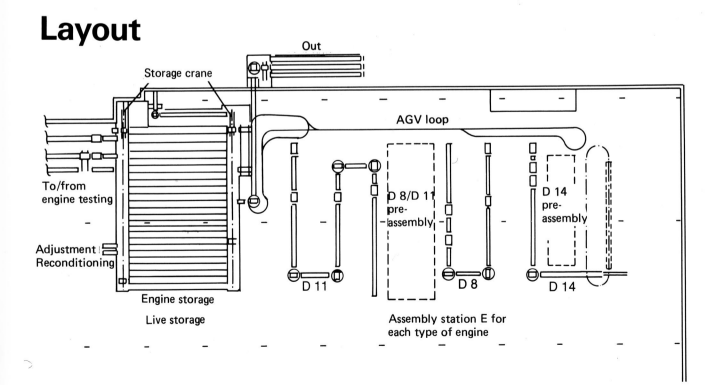

Between assembly stages, engines are moved by hand an an unpowered roller way. Assembly stations are also linked by an unpowered roller way for accumulation purposes.

The layout shows the three separate unrestricted engine assembly lines — one each for engine types D 11, D 8 and D 14. Each type is assembled on a line divided into two, or three assembly stations.

The photographs below show some of the system subfunctions.

Floor level unpowered roller way constructed from flush-mounted special sections (to reduce the possibility of operatives tripping), with walking space between each roller. The roller way assists manual movement of load carriers/assembly fixtures between assembly stations.

Automatic lifting platform for re-routing finished engines to AGVs. With the handling system chosen it has been possible to automate and integrate the transport of finished engines with the minicomputer controlled warehouse.

Automatically guided vehicle for transporting engines from assembly to storage. Automatic input of engines to warehouse line for registration, etc. AGVs were chosen instead of other fixed line restricted systems because of the available floor area and other physical aspects of the existing building. They eliminate the need for handling engines by forklift truck.

Between the assembly lines and the test department is a minicomputer controlled warehouse for approximately 700 engines, served by two cranes. To move engines from assembly to storage, a system with two automatically controlled vehicles was chosen. Finished engines are collected from automatic lifting platforms at the end of each assembly line and transported to the warehouse, where they are transferred to an input line for registering and storage.

Engines to be dispatched or to be tested/adjusted are requested by using the minicomputer. Transport is by means of roller ways, transfer trolleys and automated cranes.

The layout was designed on the basis of the available floor area, roof height, column positions, floor characteristics, etc., in the existing building, and teams who had taken part in the development work from the beginning were consulted.

The new production system has enabled station times to be extended, increasing the scope of assembly operations and leading to greater work satisfaction.

Dividing the operation into three separate assembly lines meant that workers became familiar with one type of engine and were thus able to achieve even higher quality standards. Reduced absenteeism, greater planning opportunities and better adherence to production schedules were three other results brought about by the new system.

Finally, the resultant layout is capable of producing 30,000 engines per year from a single shift operation.

Summary

The new diesel engine assembly system replaces the traditional arrangement and provides an unrestricted line for each of the three engine capacities (8, 11 and 14 litres).

Engine handling and transport are more efficient and more logical following the introduction of AGVs and a minicomputer controlled flowthrough warehouse for 700 engines.

The materials supply system has been improved. Transport/assembly fixtures have been redesigned to be uniform and to fit all engine types and handling systems.

The technical changes have made work more interesting and provided greater opportunities for individual operatives to influence their own working environment. This should result in:

- Better quality.
- Better planning opportunities.
- Better adherence to production schedules.
- Reduced absenteeism.

A constant problem during the implementation stage was the need for the reorganization to be carried out without interrupting engine production. Despite heavy demands on time and activity schedules, this requirement was met.

Goods reception in the engineering industry

The situation

A move to newly built factory premises involved the planning of a new goods reception area, etc., for goods delivered by road and rail, loaded on pallets or half-pallets with collars, and in containers. The new goods reception was intended to be an integral part of a materials handling system. Bought-in merchandize (about 50 per cent of the total) was to be checked before being put into storage, while materials of own manufacture were to go straight to the automated high-level warehouse. Materials in short supply, on arrival, had to be capable of being directed to production. The stocktaking and control of materials were to utilize an administrative system with on-line computer terminals, covering an average goods reception input of 250 pallets a day, most of the goods arriving before noon.

Alternative solutions

Two systems for goods reception were possible:

Alternative A Forklifts/loaders unloading lorries from ground level and placing pallet loads directly on the workshop floor where personnel perform the neces-sary checks. The loads are then transported by truck to the warehouse or production area.

Alternative B Pallets unloaded from lorries onto a system of conveyors for automatic transport to fixed operator positions and beyond.

Alternative A is a flexible system requiring low investment in equipment. Conversely, it requires large areas to meet goods storage and aisle requirements; transport personnel requirements are also considerable.

Alternative B requires greater investment but less floor area and fewer personnel. Materials control is simplified, and there are fewer truck/pedestrian intersections. There are fewer heat loss problems between the outdoor unloading area and the indoor goods reception. Finally, the work of the checkers is simplified since they have fixed work stations, possibly with stationary lifting equipment for heavy goods, etc. Lighting, computer links, etc., are also simplified. The above factors, combined with the large goods quantities involved, led to Alternative B being chosen.

System subfunction	Equipment chosen

Sector 1: Input

1 Unloading lorries under canopy from ground level.	Counterbalanced truck and loader with fork attachment.
2 Input to building. Possibility of excessive accumulation due to fluctuating deliveries. Need to maintain the temperature of goods in winter.	Incremental chain conveyor (alternatively, an accumulating powered roller conveyor could have been chosen). Accumulating capacity 50 pallets, approx.
3 Transport to main operator for identification/distribution.	Powered (accumulating) roller conveyor.

Sector 2: Inspection

4 Distribution of goods to checkers, high level store and production areas.	Rail-mounted transfer trolley and powered roller conveyor.
5 Goods checking stations (ten).	Powered reversible roller conveyors.

Sector 3: Storage

6 Transport to high-level store and production area.	Powered roller conveyors, chain conveyors.
7 Possibility of combining two half-pallet loads into a full pallet load.	Siding in the form of a powered reversible roller conveyor
8 Input and output to high-level store pick-and-place stations.	Rail-mounted transfer trolley accommodating two half-pallets or one full pallet.
9 Storage.	High-level storage with 6,800 pallet positions, two automated cranes.

Sector 4: Picking

10 Removal from high-level store to picking stations.	Rail-mounted transfer trolley and powered roller conveyor accommodating two pallets (storage pallet and transport pallet).
11 Picking production quantities from storage pallet to transport pallet (storage pallet returns to high-level store). Five picking stations.	Powered roller conveyors accommodating both storage pallet and transport pallet (ten).
12 Collecting pallet loads for production area.	Gravity operated roller conveyor.
13 Storing empty pallets for picking area.	Pallet magazine with powered roller conveyor.

Materials flow in system chosen

At the factory gate, arriving goods are reported and classified as a normal delivery or special urgent delivery. They are then unloaded under a canopy and placed on a roller conveyor which moves them into the warehouse. Normally, pallet loads are brought in via the reception accumulator to the goods reception operator; urgent goods, however, can by-pass the accumulator on a special conveyor.

The reception operator identifies pallet loads and enters the data at a computer terminal. Goods of own manufacture (not to be checked) proceed to the high-level storage zone, while urgent goods (shortage reported) are sent to the production area on a special roller conveyor.

Goods from external suppliers are sent to the inspection department: pallets are parked on a branch conveyor, to be distributed among the checking stations by transfer trolley. At the reception, each pallet is also fitted with inspection devices required for its contents. Three checking stations have additional accumulating capacity in the form of a separate roller conveyor. Inspected pallets are returned by transfer trolley to the line leading to the high-level store.

Pallet loads on their way to the high-level store pass an operator who identifies and labels them, and checks for weight and profile. Substandard pallets are removed from the system to a special conveyor, while accepted pallets continue towards the high-level installation. En route, individual half-pallets may be parked on a siding until a second half-pallet arrives; they are then mated and continue their journey as a full pallet load.

Transfer trolley with roller table parks pallets on inspection lines.

Input to the warehouse pick-and-place stations is by another transfer trolley which deposits the input pallet and picks up an output pallet. An output pallet requested as an intact unit goes straight to production; pallets with smaller orders go to a picking zone. Distribution there is by a third transfer trolley which has also collected an empty pallet so that the operator always receives both a storage pallet and a transport pallet. Picking completed, the transfer trolley returns and the transport pallet is placed on a line to production while the storage pallet returns to storage. Even here, there are sidings where half-pallets may be united.

Pallet en route to a transfer trolley which takes it to the high-level store pick-and-place stations. The conveyor installation consists of powered roller and chain conveyors. The latter have four chains to accommodate both pallets and half-pallets.

Layout

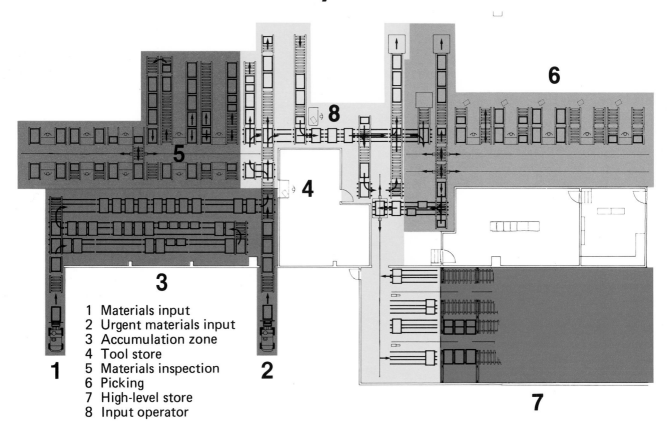

1 Materials input
2 Urgent materials input
3 Accumulation zone
4 Tool store
5 Materials inspection
6 Picking
7 High-level store
8 Input operator

Summary

The system has met the requirements laid down for capacity, versatility and environment. The next stage in the reorganization will involve integrating transport with the production area through the replacement of existing reach trucks with automatically guided vehicles.

Picking and sorting merchandise

NK-Åhléns' central warehouse at Jordbro operates as a retail warehouse in that it can supply goods direct to the group's retail outlets.

Merchandise is distributed mainly in plywood bins to the sales counters.

Incoming orders are put together from several storage areas. This means that bins destined for the same sales outlet must be collated to form unit loads.

The situation

An order picking storage installation was planned with about 3,000 picking locations and 13,000 accumulating locations for pallet loads, covering a 150 × 88 metre area in an existing building. The installation had to be capable of accommodating wide variations of operation.

Goods supply to the order picking section had to be arranged so that arriving goods, as far as possible, were handled and intermediately stored in their suppliers' original packaging. The design of the order picking storage installation had to allow for the possibility of attaching price labels to items as they were picked.

A sorting facility was to be incorporated, making it possible to sort bins to 200 destinations and having a capacity of 1,200 bins per hour.

The production system was to be reorganized without interrupting operations (this applied to the sorting facility).

The diagram indicates the various subfunctions forming part of the total materials flow The type of equipment selected for each subfunction is indicated.

Subfunctions   **Equipment chosen**

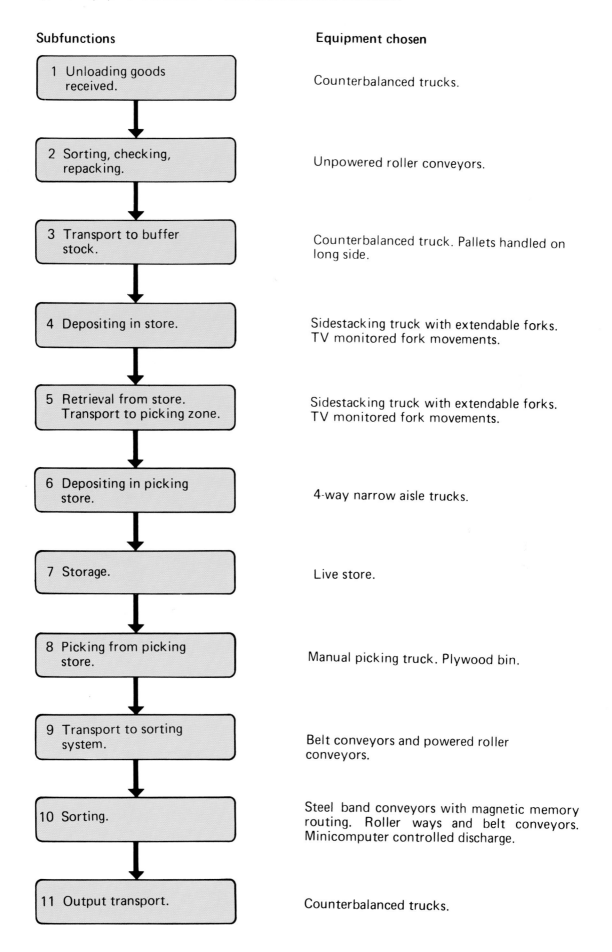

Subfunctions	Equipment chosen
1 Unloading goods received.	Counterbalanced trucks.
2 Sorting, checking, repacking.	Unpowered roller conveyors.
3 Transport to buffer stock.	Counterbalanced truck. Pallets handled on long side.
4 Depositing in store.	Sidestacking truck with extendable forks. TV monitored fork movements.
5 Retrieval from store. Transport to picking zone.	Sidestacking truck with extendable forks. TV monitored fork movements.
6 Depositing in picking store.	4-way narrow aisle trucks.
7 Storage.	Live store.
8 Picking from picking store.	Manual picking truck. Plywood bin.
9 Transport to sorting system.	Belt conveyors and powered roller conveyors.
10 Sorting.	Steel band conveyors with magnetic memory routing. Roller ways and belt conveyors. Minicomputer controlled discharge.
11 Output transport.	Counterbalanced trucks.

For the order picking section (subfunctions 6—8), the following four main types of solution were studied:

A Conventional pallet warehouse with back-to-back racking and joint wide input and output aisles, for picking at two levels.

B Pallet warehouse with separate input and output aisles. Picking section with single mezzanine. Picking from same racking at two levels. Double racking buffer store.

C Order picking store divided into two parts: a picking section designed for three-level live storage and a buffer section, with narrow aisles, for pallets in double racking.

D Picking store designed for simple block stacking on the floor, approximately three pallets deep. Picking from the outer pallet.

These four main alternatives and several variants were analyzed relative to their economic benefits and to the stated requirements. Cost of the alternatives are shown in the table below:

	A	B	C	D
Investment	1,28	1,40	1.17	1.0
Annual cost, including depreciation	0.85	0.85	0.73	1.0

To determine the best alternative, the pay-off method was used combined with an assessment of flexibility and working environment.

On the basis of these criteria Alternative C was chosen. It is described below.

Live store order picking

Reception area
Arriving goods are unloaded from trucks or containers by battery powered counterbalanced trucks.

Unloaded goods, checked against documents, are notified via a terminal directly linked to a central computer.

The computer produces unpacking documents for the notified consignment. With the aid of these documents goods are sorted to produce pallet loads of specific items. Quantities are randomly checked during sorting. The size of suppliers' packages is checked against data in the unpacking documents. Any variation results in the consignment being sent to a repacking station where it is reduced to a suitable size for storage.

Sorted and checked pallets are labelled for removal to vacant buffer storage positions or, where necessary, straight to the picking zone.

Transport between goods arrival and buffer picking is by conventional battery powered counterbalanced trucks.

Accumulation and picking zone
The buffer zone has conventional pallet racking, about 7 metres high, for about 12,000 Euro pallets. There are five levels of racking at 1.65 metre centres (see figure).

The racking is connected directly to three levels of picking mezzanines on each side of which are 21 aisles with a free width, at maximum load profile, of 1650 mm.

The picking zone has a semi-live storage installation with three picking levels. At each level there are approximately 1,000 roller ways, 6 metres long, and of various widths (300, 400, 500 and 600 mm). The roller ways are located at four heights at each level (20, 70, 110 and 150 cm above their base). Complete suppliers' packages are placed on the roller way from the buffer store or straight from goods arrival.

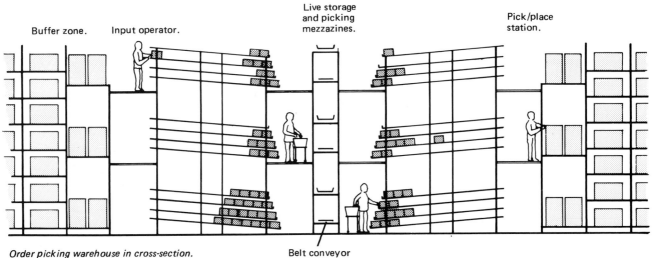

Buffer zone. Input operator. Live storage and picking mezzazines. Pick/place station.

Order picking warehouse in cross-section. Belt conveyor for picked goods

Items in live storage are selected with the aid of picking orders produced by the computer.

A picking order consists of a number of sections with self-adhesive address and price labels printed in the order in which the items are stored. For each item ordered there is a 'start label' giving item number and storage location, and the required quantity of price labels.

Live storage picking zone with central belt conveyor to remove bins containing picked items.

Moving along the live storage compartments, the picker stops at the compartment indicated by the 'Start labels', opens the supplier's package, and picks and prices the number of items corresponding to the number of price labels. The goods are placed in bins (one shop per bin). Picking completed, the bins are transported by conveyor to a sorting system where they are combined into unit loads for individual shops.

The picking zone is served by four high-lift narrow aisle trucks equipped with transverse extendable forks. Fork movements are monitored by a TV camera linked to a monitor display near the driver's seat. The trucks operate in the accumulation aisles between reception terminals at one end and input terminals at the picking mezzanine end (one terminal on each level). Between input terminals and live storage compartments, special operators using input vehicles see to it that goods end up on the correct roller ways.

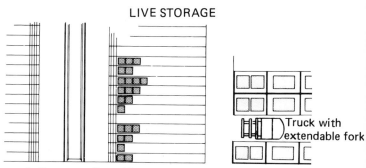

LIVE STORAGE

Truck with extendable fork

Buffer store pallet racking is served by trucks with extendable forks.

A narrow aisle truck has moved a pallet to the buffer store pick-and-place station where the input operator follows up using a modified fork carriage.

Sorting system
Goods for individual sales outlets are picked in bins in various zones in the warehouse. To combine these into convenient output loads some kind of sorting system is required.

Previously, the central warehouse at Jordbro had a sorting installation with 36 address lines and a capacity of about 800 bins per hour. It served only a few warehouse zones. The expected increase in goods volumes and the desire for uniform sorting of bins from all zones suggested an installation capable of sorting binned goods to 200 addresses, with a capacity of 1,200 bins per hour.

The choice was to revamp the old system or to design something completely new. The latter alternative was chosen.

The sorting system consists of several belt and roller conveyors from each storage zone. The various bin flows are merged via transfer switches, flowing to three stations where plywood lids are fitted by hand, with subsequent automatic strapping at special stations.

Bins flow together from the conveyors and arrive at the three strapping stations in roughly equal proportions. The entire flow can be directed to one, two or three strapping stations with a manual transfer switch. After lidding and strapping, goods rejoin a single flow which passes a labelling station equipped with a Sandvik magnetic memory system (MMS).

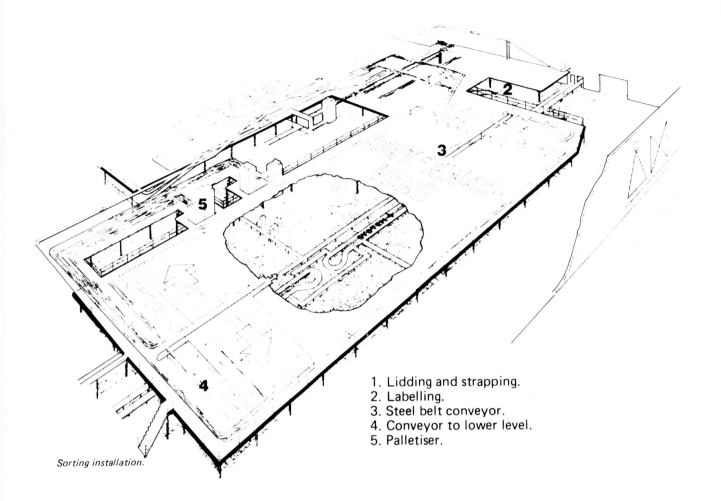

1. Lidding and strapping.
2. Labelling.
3. Steel belt conveyor.
4. Conveyor to lower level.
5. Palletiser.

Sorting installation.

A bin has just joined the steel belt conveyor.

Bins are fed one by one by a pusher to a steel belt conveyor. The belt is magnetized with a code corresponding to the bin addresses and derived from the bin label via a scanner at the loading end. At right angles to the steel belt are 200 roller conveyors (100 along its top run, 100 along its return run). Bins are transferred to these (sorting) conveyors by L-shaped pushers (one for each conveyor). Each sales outlet has its own conveyor address. A reader at each roller conveyor checks the magnetic code and, when it corresponds to the reading code for that conveyor, the pusher removes the bin from the steel belt.

Reloading bins from the top conveyor section to the return section is also done by an L-shaped pusher in combination with a descending roller conveyor loop.

Each sorting conveyor can accommodate ten bins. If a conveyor is full, a bin will by-pass it and return to the labelling station for re-coding. Sorting lines are emptied automatically by a microcomputer programmed to act when a line has accumulated a pre-set number of bins. The 'bin train' from this line moves to a palletiser. The resultant pallet load is strapped automatically before being removed to the dispatch bay.

Bins queueing on the sorting lines (unpowered roller conveyors) for transfer to the palletiser.

Summary

The new order picking warehouse provides more efficient and logical picking of retail items. In addition, the environment has been improved in that access to goods for picking is easier and input and output operations, which previously involved collision risks, are now separate.

Moving pallet loads in a furniture warehouse

Automatically guided vehicles move goods to the crane served racking via an ID station.

The situation

A furniture distributor expected a growing demand from the domestic market. Old buildings and the existing high-level stacking warehouse with 25,000 m² of floor area and a storage height of 7.5 metres, served by narrow aisle trucks, were scheduled to be extended. A new 30,000 m³ capacity building was to be positioned adjacent to the existing warehouse.

Goods arrived by rail or road to be dispatched primarily by road. In the existing warehouse goods were stored on Euro pallets and large in-house standard pallets. Goods arriving as loose items were loaded on pallets.

The administration system based on an IBM 370 computer was well developed.

Alternative systems

Three handling alternatives were examined:

- Conventional reach trucks.
- Narrow aisle trucks.
- Automated high-level store.

Following an analysis of building costs, layout storage techniques and future expansion needs, and communication between units, the company chose automated high-level storage.

The diagram shows the system subfunctions and the equipment selected.

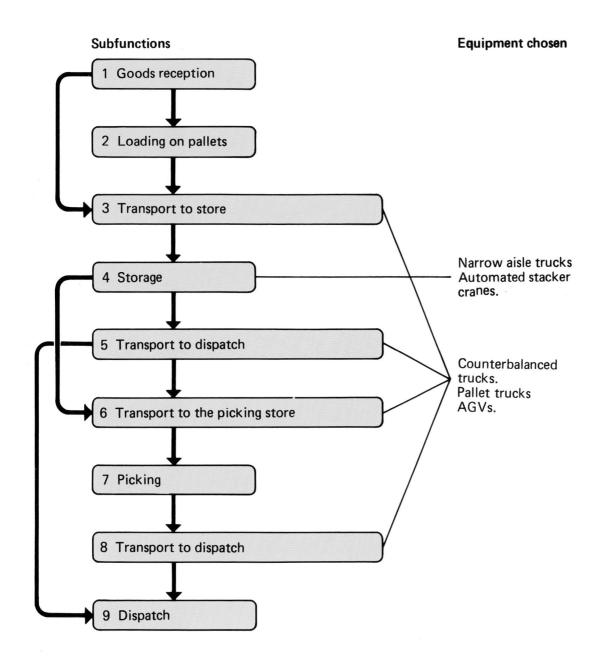

Subfunctions	Equipment chosen
1 Goods reception	
2 Loading on pallets	
3 Transport to store	Narrow aisle trucks Automated stacker cranes.
4 Storage	
5 Transport to dispatch	Counterbalanced trucks. Pallet trucks AGVs.
6 Transport to the picking store	
7 Picking	
8 Transport to dispatch	
9 Dispatch	

Solution

At goods reception, goods are sorted and labelled for transport to high-level racking or the automated store. Transport to a powered roller conveyor is handled by counterbalanced trucks.

The roller conveyor is equipped for checking load profiles, weights and pallet quality. Rejected pallets are returned for further action. The automated store can accommodate pallets with maximum plan dimensions of 1,400 × 900 mm and heights up to 1,150 mm. Overhang is acceptable for 1,200 × 800 mm pallets — a novelty in the context of automated warehouses. While being checked, pallets are also centred.

Pallets are transferred from the roller conveyor accumulating section to AGVs which pass an ID station: here, pallet item numbers are transferred to the computer by an operator using a pen reader. The AGVs then move the pallets to an alignment device before stacker cranes collect and transport each pallet to its designated location.

Pallet output operations use the reverse working order, excluding identification, alignment and checking. At the other end of the conveyor line, an automatic printer produces address labels. These are taken by bay area truck drivers, each label being delivered with its corresponding pallet load to the specified dispatch location.

Handling equipment

The bay area is served by counterbalanced trucks and has roller conveyors with checking stations.

The high-level stacking area is served by three narrow aisle trucks.

Narrow aisle truck.

Thirteen 700 kg capacity AGVs provide the link with the store. They are induction loop controlled and battery powered. When not operating, they return to their parking places where their batteries are recharged. The vehicles have a speed of 1.1 m/sec. which is automatically reduced to half speed or crawling speed when required.

Parked AGVs.

The automated store is equipped with 10 stacker cranes operating in 186 metre long, 16 metre high aisles. Each crane is controlled by a microcomputer receiving data at the end of the aisle from a main computer. Data are transferred by infrared light. The cranes are approximately positioned by a pulse generator and fine positioned by photoelectric cells sensing reflecting tape on the racking. Travel speed is 150 m/min and lifting speed 30 m/min.

AGV placing goods for a stacker crane.

Investment

The cost of building, site improvement, bays and equipment, including a co-ordinating minicomputer, was around SEK 35 million.

Specific warehouse facts

- Floor area: 200,000 m². Capacity: 1 million m³, approx.
- The entire central warehouse employs 370 people, 290 of whom are operatives and 80 salaried staff.
- About 60 long-distance lorries and railway wagons arrive daily to be unloaded. This amounts to nearly 4,000 m³ of goods
- Each week 425,000 items (equivalent to 40,000 m³) are handled. To aid operations there are 180 powered vehicles and handling devices.

Order picking using trucks

Pallet truck used for order picking.

Order picking and compilation represent a theme running through all distributive activities. Different situations lead to solutions based on different principles. In choosing methods, premises sometimes create limitations, sometimes possibilities. Often old buildings, not entirely suitable for the purpose, must be used, limiting the choice to a few alternatives only. On other occasions new building plans provide complete freedom of choice.

This section describes two solutions: one where old premises were used, the other where a building was purpose-built.

277

Reorganisation of storage and order picking on existing premises

The situation

The warehouse for finished products at ASSA-Stenman in Eskilstuna was to be reorganized, the picking/packing to be rationalized. Simultaneously, the product range (locks, lock fittings and forgings) was to be reduced and the remaining volume expanded.

Existing premises were to be used. Ceiling heights vary somewhat between 4 and 4.5 metres, and headroom is reduced by transverse girders. The floor has great load-bearing capacity. Columns, walls and gates impose limits.

Existing sheet metal shelving was basically in good condition and was to be reused.

High-lift order picker.

Subfunctions	Equipment chosen
1 Storage.	Shelving. Pallet racking.
2 Order picking.	Free path high-lift order pickers.
3 Transport to packing.	Free path high-lift order pickers.
4 Packing	Order pickers and special packing tables.
5 Dispatch.	Hand pallet trucks.

The problem of picking — a key function under existing conditions — was solved by the use of high-lift order pickers. These have a capacity of 1,000 kg and are provided with an adjustable lift feature to ensure that the fork height provides a comfortable working position for the driver. The driver can be elevated so that he can reach items stored 4.5 metres above floor level. Limiting devices prevent collision with roof girders.

The distance between shelves is 1,400 mm. Order pickers carry a standard pallet crosswise and need a minimum 1,200 mm free aisle width. Guide tracks on the floor assist driving in the relatively confined space.

All items have a special location and there is no buffer store. Drivers must, therefore, be able to reach the entire range of stored items.

Items requiring most space are stored on standard pallets in racking, not on shelving.

Order lists indicate items in picking order. An average order covers 20 locations, and the average package weight is 1 kg. However, order sizes vary greatly.

Picking is based mainly on the pick-and-place principle, goods being picked from the shelving and put directly into packaging on standard pallets.

Goods are moved to the dispatch zone, in some cases to a special packing zone, by order pickers.

To handle the approximately 50,000 orders (total weight 5,000 tonnes) annually, six free path high-lift order pickers are needed.

Order picking groceries in regional wholesale warehouse

The situation

A regional warehouse belonging to a large firm of wholesale food merchants was to be built and current operations transferred from the old premises. The range of merchandise is divided into groups, one of which — groceries — is dealt with here. The number of items, handled volumes and handling frequencies are known. Goods were to be stored indoors at normal temperatures, and no special storage or handling arrangements were needed. Standard pallets were to be used as load carriers, for external transport as well as for storage, with roll pallets carrying output goods.

Orders were to be picked on the basis of computer arranged picking lists. Volume and weight of individual orders vary. Average item weight is 10 kg. The merchandise was to be stored in a manner which reduces picking.

Subfunctions   **Equipment chosen**

1 Goods reception.	Pallet trucks. Hand pallet trucks.
2 Transport and input to store.	Reach trucks.
3 Storage.	Pallet racking or free stacking.
4 Transport from accumulation point to picking point.	Reach truck or ride-on stacker.
5 Storage.	Pallet racking or free stacking.
6 Picking.	Manual picking. Low-lift order pickers.
7 Transport to dispatch zone.	Low-lift order pickers.
8 Dispatch.	Roll pallets.

Layout

The warehouse is rectangular, and goods flow is from reception at one end to dispatch at the other. At one side, installations are permanent; any expansion would be accommodated on the opposite side.

Warehouse pallet racking stretches in the direction of the goods flow. Situated next to the output loading zone is a free-standing pallet area for items requiring most space and/or those most frequently handled.

The 45 metre long racking modules stand back to back, with 2.5 metre wide aisles. They permit four pallet load levels at 1.25 metre centres, the two bottom levels being used for picking and the top levels for buffer storage.

Each item has a fixed pallet location but buffer store locations may vary.

Transverse aisles enable the loop to be shortened and, possibly, items to be accessed more easily.

Order picking equipment

The 13 picking operatives all use pallet trucks or low-lift order pickers.

Forks have been lengthened to take three roll pallets or two standard pallets. In addition, vehicles are fitted with desks for the picking lists.

For reasons of ergonomics and greater speed, pallet trucks are gradually being replaced by low-lift order pickers.

Picking

The picker collects empty roll pallets from the dispatch zone and then follows the normal picking loop, picking items in the designated order and attaching pre-printed labels to each item.

Filled roll pallets are moved to the dispatch zone by the picker who then takes on board empty roll pallets and continues to pick.

When items in picking locations run out, the picker calls for a pallet load to be brought down from the buffer store.

On average, each picker handles 175—200 goods items per hour. This time includes collecting empty roll pallets, removal of picked goods and waiting for buffer stored loads to be produced.

Low-lift order picker with accommodation for three roll pallets.

Pallet truck accommodating two standard pallets.

SKF's Gothenburg warehouse for the Scandinavian region

Stacker cranes parked at the end of the aisles.

The situation

To rationalize production, SKF split up its vast product range between individual factories. About one-third of the range required by the Scandinavian market is produced at the Gothenburg factory, while the other two-thirds is imported. In turn, the factories in Britain, France and West Germany obtain from Gothenburg products they themselves do not produce.

Increased in-house distribution meant that the warehouse installation which had previously been shared for distribution in Scandinavia and other European countries was no longer large enough.

Distribution patterns for the two regions differ considerably. While exports to Europe (including Britain) are in large consignments, the Scandinavian distribution consists mainly of small consignments direct to dealers and users. It was decided, therefore, to separate Scandinavian distribution from the existing organization and accommodate its special requirements in new premises designed for good space utilization.

Order patterns

The Scandinavian regional warehouse is designed solely to meet the needs of a local market, distributing large numbers of products in small volumes which involves much order picking. The need for quick and reliable deliveries is great. For instance, it should be possible for goods to reach their Scandinavian destinations according to the following priorities:

24 hours for goods (one-third) dispatched to users in Sweden;

48 hours for goods (one-third) dispatched to district depots in Sweden and for goods (one-third) dispatched to sales depots in other Scandinavian countries.

The number of items, including packaging variants, is about 10,000, consisting of bearings of every conceivable dimension up to sizes which need to be handled on half-pallets.

Subfunctions	Equipment chosen
1 Transport to picking zone.	Automated conveyor system.
2 Picking for input store.	Stacker crane with extendable forks, equipped for picking.
3 Storage	Brackets for half-pallets.
4 Picking.	Stacker crane with extendable forks, equipped for picking.
5 Transport from picking zone.	Automated conveyor system.

The installation was designed to dispatch approximately 5,000 order lines per week and to receive approximately 500 deliveries in the same time.

At present (1980), about 600 order lines per day are dispatched, corresponding to about 750 picked items and a total goods weight of 28–35 tonnes. About 10 per cent are urgent orders.

Choice of handling method

Of the total goods output only about 15 per cent is dispatched as complete half-pallet loads. This simplified the choice of picking method (station picking or rack picking). Goods quantities picked are so small that a pallet must be picked 8–10 times on an average before it is emptied. Handling costs of station picking would then be unreasonably high; therefore, picking from racks was chosen.

Equipment and building heights were chosen partly as a result of a comparative study of building costs and partly on site availability.

Building costs per unit of area are affected comparatively little by increased building height. Increased costs of handling equipment for high-level racking are more than offset by the reduced costs of constructing a more compact building. Also, the available site need not be utilized fully, offering potential for later extensions.

Ground conditions are very poor and extensive piledriving was required. Maximum pile depth is approximately 60 metres, average depth approximately 42 metres.

The shape and weight of the goods make them suitable for storing on 800 × 600 mm half-pallets fitted with two collars. Smaller quantities are placed in metal bins which are also stored on half-pallets.

If pallets were placed in conventionally spaced racking, their low dimensions would waste usable height. It was decided, therefore, to use special brackets. The entire space between the top of one pallet and the bottom of the next can thus be utilized for picking. This technique enables 12 pallets to be stacked vertically in the 10 metre high warehouse.

Picking could have been from driver-elevating order pickers or from stacker cranes with picking equipment. The former offered simple aisle changes and would thus have allowed one truck to pick complete orders from several aisles. Against this was the risk of queues forming in the aisles, with the further disadvantage that, in view of the working heights involved, goods input would have demanded other trucks, thus exacerbating the queue problem.

Cranes, on the other hand, permitted input/output operations from one machine as well as speedy transport, access to all pallets stored and operator safety. Some orders could not be picked by one and the same crane but this drawback was overcome by moving partly filled pallets on conveyors between the aisles.

Layout

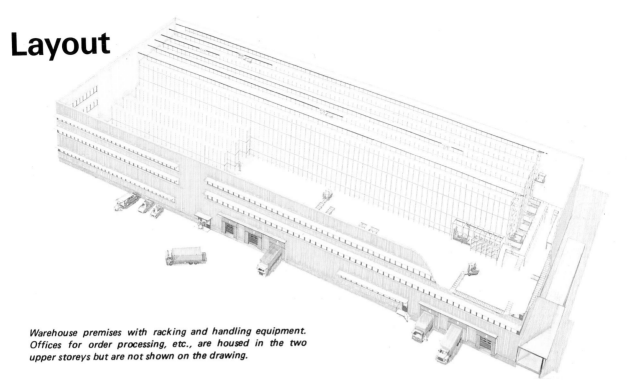

Warehouse premises with racking and handling equipment. Offices for order processing, etc., are housed in the two upper storeys but are not shown on the drawing.

The Scandinavian regional warehouse measures 40 X 90 metres and roof height is 10 metres. The actual store consists of pallet picking racking 82 metres long in 10 metre high aisles.

Goods transport to and from the picking zone and between the aisles is by automatic conveyor. The racking in which beams have been replaced by brackets can accommodate 12 pallets stacked vertically. Total storage capacity is 13,500 half-pallet loads and approximately 8,000 smaller containers (placed on half-pallets). In one of the aisles, two pallets are stored one behind the other in the racking. In these locations a standard pallet may replace two half-pallets, if required.

The six order picking cranes have a maximum travel speed of 150 m/min and maximum lifting speed of 30 m/min. They are equipped to pick small items and, using their extendable forks, full pallet loads. Special equipment for approximate positioning guides them automatically to the position selected. A display indicates continuously the crane's position to the driver.

When picking, the driver normally takes goods directly from a pallet. If items are heavy, he can lift out the entire pallet load and place it next to his receiving pallet on the crane. Picking can then be carried out conveniently, after which the storage pallet is returned to the racking.

The warehouse is equipped with sprinklers throughout. Extra protection is provided by sprinklers at three levels in the high-level racking.

Aisles are 82 metres long and 10 metres high.

Order processing

When a customer places an order, it is recorded on a VDU terminal. If delivery by return is required, all documents such as order confirmation, invoice and delivery note are printed simultaneously. The delivery date is stated *en clair* on the screen. The order picking list then goes to the picking zone for implementation. After picking, the order is checked and packed for transport.

Flexibility

The various warehouse functions experience greatly varying work loads. Incoming orders increase and decrease periodically, and goods arrivals are irregular. The problem has been solved by training each employee to handle almost any of the warehouse tasks. This means there are 11 drivers available for 6 cranes. Also, computer terminal operations can be performed by warehouse operatives. Good team spirit means that personnel from sections which are less busy help out elsewhere when required.

Investment costs

Investment in the building inclusive of offices amounted to about SEK 12.4 million, SEK 2 million of which went on the foundations. Warehouse equipment cost approximately SEK 6 million.

Index